MOUNTAIN FLOWERS

Text by Harvey Manning
Photos by Bob and Ira Spring

The Mountaineers • Seattle

The Mountaineers: Organized 1906 ". . . to explore, study, preserve and enjoy the natural beauty of the Northwest."

Published by The Mountaineers, 306 2nd Ave. W., Seattle, WA 98119

Distributed in Canada by Douglas & McIntyre, Ltd., 1615 Venables St., Vancouver, B.C. V5L 2H1

Design by Marge Mueller. Front cover: Meadow flowers and Sloan Peak. Back cover: Avalanche lilies on Mazama Ridge, Mount Rainier National Park

Library of Congress Cataloging in Publication Data

Manning, Harvey.
Mountain flowers.

Bibliography: p.
Includes index.
1. Wild flowers—Cascade Range—Identification. 2. Wild flowers—Washington (State)—Olympic Mountains—Identification. 3. Alpine flora—Cascade Range—Identification. 4. Alpine flora—Washington (State)—Olympic Mountains—identification. I. Spring, Robert, 1918- joint author. II. Spring, Ira, joint author. III. Title.
QK192.M36 582'.13'09795 79-9284
ISBN 0-916890-92-9

INTRODUCTION

A friend of mine, the late Weldon Heald, once went hiking in the High Sierra with an old man of the mountains, the sort who knew every crystal in the rocks and song in the forest, the warning of a wispy cloud and the drama behind a track in the mud, and the names, English and Latin and Indian, of all the trees and shrubs and herbs and mosses and lichens. It was better than a college education, walking with the professor, each day absorbing lore gleaned by the master over decades. And it came to pass that Weldon, as often before on the trip, found an unfamiliar flower, a small blue blossom, and as often before, and never disappointed, asked his learned companion. Mountain man looked, sucked air, swelled up halfway to apoplexy, and exploded, "Oh, that's that gollydang little blue flower!"

As it was once with the old man of the mountains, so it is constantly with novice scholars. The path to wisdom is strewn with gollydangs. In this booklet are photographs and descriptions to help a beginner put names on 84 of them. Some 150 others — relatives, look-alikes, and frequent associates — are mentioned to guide further study elsewhere.

On a summer walk in the Cascades or Olympics, climbing from low valleys to high ridges, the casual observer may see 125-150 flowers, though on an exceptionally diverse route the number can rise near 200. The more one learns the more one sees: the blur of "meadow white" distills into cow parsnip, meadow parsley, Sitka valerian, bistort, pearly everlasting, yarrow, sandwort, catchfly — and assorted little white gollydangs.

Explanation of the Descriptions

The heading of each flower description starts with common names, first the preference of the author, then others if there are any. There nearly always are. In one valley a certain plant comes to be called tow-head baby, and up the creek a ways, dishmop, and beyond the ridge, old man of the mountains, and in the next county, mouse-on-a-stick; tourists from the city call it mountain pasqueflower, and from the next state, western anemone. Attempts to standardize common names have failed. A person is free to call a flower whatever name he pleases, and if he doesn't know any to invent his own, such as "little blue gollydang."

The freedom doesn't extend to the scientific, Latin names; these are standard and universal — except for changes as botanists learn more and refine, or argue about, classifications. The primary subdivision of the plant world is families; native to Washington are about 100 containing flowering trees, shrubs, and herbs; 28 are represented here. A family is broken into genera (the plural of genus); the Arum (or Calla Lily) Family is represented in Washington by just a single genus, the Sunflower (or Aster, or Composite) Family by half a hundred. A genus is broken into species; in Washington mountains and foothills are roughly a thousand flowering species. A species may be broken into subspecies, also called races, but most widely (and in these pages) varieties; the differences may be invisible to the untrained eye or so major that the untrained eye marvels, as in the case of the magenta and white paintbrush, that experts consider them the same species.

The scientific name, always italicized, has two parts: the first, always capitalized, is the genus; the second, never capitalized, is the modifying adjective that defines the species. The first can be used alone without confusion because no two genera have the same name; a student may identify a plant only to the genus and be well pleased; it is something of a victory, after all, to distinguish a *Castilleja* from a *Sedum.* The second term cannot be used alone because such adjectives as *uniflora* (single flower), *formosa* (beautiful), and *tolmiei* (honoring Dr. Tolmie) are used to define very unlike species in totally different genera. Were Weldon's wilderness professor to systematize his ignorance in a pseudo-scientific manner, lumping all unknowns in a genus, then characterizing each by adjectives, he might call his nemesis *Gollydang littleblue.* Except he'd have to translate that into Latin.

Flowering season. The spans given here are very approximate, merely suggestive. A plant that grows from sealevel to highlands flowers from low to high over months, but only a few weeks at any one elevation. Additionally, the bloom is earlier on a warm south slope than a cold north, on a sunny knoll than in a shadowed ravine. And when snowbanks linger deep and late the flowering climax usual in subalpine meadows in late July to early August may be delayed to September.

Features. In a look-and-match book such as this the photographs carry the main burden. Features are described in the text either because they are not plainly shown in the picture, as a leaf hidden by the blossom or a perfume unrecorded by the camera, or because they are of particular help in identification. The approximate maximum height is given — with the warning that in many species the range is from low-huddling inches to high-thrusting feet. In order not to impede a beginner by endlessly sending him to the glossary, none is provided and the language is nontechnical, which means inexact; for the precision that is the soul of science the student must advance to other works.

Habitat. Some plants adapt equally well to dry sites and wet, but most prefer one or the other, or forest shadows or meadow sunlight, or rich leaf-mold or sterile gravel or cracks in rock walls; the likings are noted here, with the caution that many species will take whatever they can get (after having been driven out of the choice spots by competitors) and be happy.

A plant's elevation range varies considerably over the latitude range from southern to northern Washington, as well as over the climatic range from western to eastern Washington — as well as from one side of a ridge to the other. Again sacrificing precision, the concept of vegetation zones basic to science is left to other books and a more primitive terminology used here. "Lowlands" extend from sealevel some distance into mountain valleys; the old-growth forest in most Western Washington lowlands is dominated by western hemlock, Douglas fir, and western red-cedar. "Middle elevations" are still deep woods except where broken by lakes and marshes and avalanche swaths and such; Pacific silver fir and Alaska cedar are dominant trees; this is snow country, the winter white commencing in late fall and remaining well into spring. "Subalpine" covers the transition from continuous forest to open meadow. The two zones interfinger: trees (characteristically short and pointy-topped, among the dominants being mountain hemlock, subalpine fir, Engelmann spruce, whitebark pine, or larch, depending on the place) follow ridge crests upward and crown rocky knolls; meadows (typically of the lush, "tall-grass" kind, knee-high or more) carpet cold basins where snow stays late, follow gullies downward, and sprawl over boggy flats. This is the parkland the layman usually calls "alpine," but botanists restrict that term to the land above the trees, the "short-grass" meadows, the tundra country of high ridges.

Distribution. Most flowers pictured here grow in both the Olympics and Cascades. Some are confined to one range or the other, or to the east or west side, or north or south end, of the Cascades.

Though this booklet focuses on these two mountain ranges of Washington, it is useful in southern British Columbia and northern Oregon. Dramatic as the influence of latitude is when tropics and arctic are compared, from one state or province to the next the differences in flora are just barely noticeable by the beginner. The wide gap of the Columbia River is a barrier to reckon with but over the eons has been more of a problem for people than plants.

Associates. For most flowers one or several common comrades are listed to suggest the company they keep. The associations noted here are of time as well as place: though glacier lily and spring beauty share ground with a host of associates that blossom later in summer, by then the two early bloomers long since have gone to seed.

Botanists use the term "association" to denote a very specific community. For example, in the Pacific Silver Fir Zone (the vegetation zone that covers most of what here is called "middle elevations"), some five associations have been described, including the Pacific Silver Fir/Alaska Huckleberry Association where the forest-floor cover is typified by Alaska huckleberry, oval-leaf huckleberry, Canadian dogwood, queen's cup, and twinflower. In subalpine meadows of Mount Rainier five major type groups of plant associates are recognized, including the Heath-Shrub Community dominated by red heather, white heather, and Cascade huckleberry, and the Lush Herbaceous Community dominated by Sitka valerian, green hellebore, subalpine lupine, magenta paintbrush, glacier lily, avalanche lily, western anemone, mountain bistort, fireweed, and cow parsnip.

Early on in his mountain walking a person entering a meadow views it as a mob of gollydangs united in a mass color riot. The first step toward comprehension is breaking each color into species. The next is putting the species together again in a community. The riot is then the more admired for being seen not as senseless anarchy but an orderly, concerted plot against a hiker's reason.

Suggestions for Further Study

A large and rich literature can be but sampled here, suggestions limited to books readily available in well-stocked bookstores and most pertinent to needs of a primary student — with notations for the time when he progresses to secondary school.

The definitive work on vegetation zones and plant associations and communities is *Natural Vegetation of Oregon and Washington,* by Jerry F. Franklin and C. T. Dyrness (U.S. Forest Service, Pacific Northwest Forest and Range Experiment Station). The person who digests the entirety rates a bachelor's degree at least, maybe a master's, yet the book is easily readable by an interested layman and there's no other volume so comprehensive and good. The most dependable places to buy it are the Government Printing Office Bookstores in Portland and Seattle.

Land Above the Trees: A Guide to American Alpine Tundra, by Ann H. Zwinger and Beatrice E. Willard (Harper and Row), is a layman's introduction, soundly scientific but with poetics appropriate to turf and snowbed, fellfield and felsenmeer, gopher garden and krummholz, the realm where "winter is never more than six weeks away."

Though the flowers pictured here account for maybe two-thirds of the color along a mountain trail, the remainder is omitted for lack of space. Proper representation has been denied the enormous Sunflower Family, also called the Composite Family because typically its heads are composites of two different kinds of flower, the outer ring of ray flowers that the layman calls "petals" because they are like the true petals of other families, and disc flowers in the center; the daisy is a familiar example. Scores of "daisies" and "sunflowers" and "dandelions" line the trails — but actually are things like balsamroot and hawkweed and agoseris that by their repetition of form and color tend to bore photographers. Absent here altogether is the large Mustard Family (20 genera in Washington), the *Cruciferae* whose flowers have four petals forming a cross. Flower-book photographers are prejudiced against the family because with few exceptions, notably the brilliant yellow wallflower of high and dry ridges, its blossoms are small and hard to photograph. Flower-book authors are prejudiced against the family because many of its teensy-weensies, such as rock cresses and whitlow grasses, are very hard to describe or even tell apart; the family abounds in little blue (and purple and white and yellow) gollydangs. In the Pink Family, no room has been found for the sleepy catchfly or sandwort, nor in the Waterleaf Family for the pale-blue ballhead waterleaf so dazzling on springtime trails or the stunning silky phacelia of alpine ridges. And in the Lily Family the hiker must use his unaided nose to find the onions.

Remedies for deficiencies of this volume can be found in other look-and-match books offering color photographs and concise, layman-readable descriptions. *Wildflowers of the Olympics,* by Charles Stewart (Nature Education Enterprises), has 100 photos of flowers, most of which are in the Cascades as well. *Wildflowers 1: The Cascades from Canada to California,* by Elizabeth L. Horn (Touchstone Press), has 135 flowers, most found in Washington, and in the Olympics. *Mountain Wild Flowers of the Pacific Northwest,* by Ronald J. Taylor and George W. Douglas, photos by Lee Mann and Ronald Taylor (Binford and Mort), pictures 144 flowers and features a unique key especially designed for beginners. *Wild Shrubs, Finding and Growing Your Own,* by Joy Spurr (Pacific Search Press), portrays in photos and drawings 40 flowering shrubs and tells how to bring them to a home garden. (This is ticklish business, of course. The rule that is absolute in national parks ought to be generally observed elsewhere — look all you want, and sniff, and photograph, but never pick. Still, just as there are places trees can be cut for lumber and pulp without outraging nature and the conscience, so there are suitable spots to dig plants. Or eat them.) The huge and gorgeous and costly (but worth it) *Wild Flowers of the Pacific Northwest,* by Lewis J. Clark (Gray's Publishing/Superior Publishing), has more than 600 color plates and a scholarly, delightful text discussing 800-odd species and subspecies. With photos and essays scaled down, the book has been broken into a series, *Lewis Clark's Field Guide,* consisting of six pocket booklets on *Wildflowers of Field and Slope, of Forest and Woodland, of Arid Flatlands, of Marsh and Waterway, of Sea Coast,* and *of Mountains.*

Fond, sad mention must be made of the state's pioneering look-and-match color book, now priced out of the market by inflation. Many photographs from *Wildflowers of Mount Rainier and the Cascades* (The Mountaineers) have been reused here, but for the matchless essays by Mary A. Fries, the present author's first teacher, a student must go to a library.

Trees, Shrubs, and Flowers to Know in British Columbia (or *Washington*), by C. P. Lyons (J. M. Dent), has no photos but for its superb drawings and descriptions is still, after a quarter-century, perhaps the best second-level book, covering some 43 trees, 105 shrubs, and 335 flowering herbs. *Washington Wildflowers,* by Earl J. Larrison, Grace W. Patrick, William H. Baker, and James A. Yaich (Seattle Audubon Society), has small color photos of 250 flowers and black-and-whites of others; its unique value for the intermediate scholar lies in the cataloging of 1134 species of native flowering herbs — perhaps 90 percent of the total in the state. (It perhaps needs pointing out that my text leaves the distinctions between herb and shrub, and shrub and tree, to these higher-level treatises.)

The end of the line for most nonprofessionals is *Flora of the Pacific Northwest,* by C. Leo Hitchcock and Arthur Cronquist (University of Washington Press). By the time a person can find his way around in the Big Book he'll be a far piece down the trail from wide-eyed bewilderment of look-and-match days. But still to enliven his wildland walking there will be — greeted no longer solely with frustration but now with at least as much exhilaration — many a little gollydang.

Harvey Manning

TRILLIUM
Wake robin
Trillium ovatum
Lily Family

Skunk cabbage's yellow and coltsfoot's white usually are the first of a new year's colors amid the old winter's monotones. Closely following, in the shrub layer above the forest floor, are the white of Indian plum and the red of currant. But only when the green tubes of trillium poke through brown litter and unfold leaves and petals is it officially springtime in the woods. Before the bright splashes of white fade to purple, other plants will be joining the bloom to carry on the show.

Flowering season: March to August. **Features:** three large petals, three large leaves; up to 10 inches tall. **Habitat:** moist forests, lowland to subalpine. **Distribution:** Olympics and Cascades. **Associates:** skunk cabbage, yellow violet, bleeding heart.

COLTSFOOT
Butterbur, sweet coltsfoot
Petasites frigidus
Sunflower Family

Because it pioneers gouged and trampled ground, such as roadsides and trails, the coltsfoot often is ragged and dusty by summer, the more disreputable for the scruffy balls of winged seeds awaiting winds to carry them to new frontiers. However, because it also pioneers spring, competing with skunk cabbage for honors as first bloomer, coltsfoot attracts the eye hungering for an end of winter. Viewed in the mass the flower heads may seem nicely bright but rather gross; on close study, though, each massive congregation resolves into hundreds of curled white stars with purple centerparts. Big but dainty.

Flowering season: March to June. **Features:** white to pinkish flowers in compact heads, maturing to balls of dandelionlike seeds; up to 24 inches tall. **Habitat:** disturbed sites in moist woods and meadows, lowland to subalpine. **Distribution:** Olympics and Cascades. **Associates:** skunk cabbage, salmonberry.

A small variety, alpine coltsfoot, *P. frigidus* var. *nivalis,* grows beside streams in the snow country of subalpine meadows.

VANILLA LEAF
May leaf, deerfoot, sweet-after-death
Achlys triphylla
Barberry Family

Mists (that's what *achlys* means) of white bloom float above a green carpet. If a hiker crumples a plant in his shirt pocket the sweet scent veils the reek of days on the trail. Frontier women putting away new-washed clothes sprinkled them with the leaves to soften the harsh smell of homemade soap. They also hung clusters from walls to perfume cabins (and, oddly, to repel flies).

Flowering season: April to July. **Features:** three large leaves, usually about 1 foot from ground, often forming carpets. **Habitat:** deep woods from lowlands to middle elevations. **Distribution:** Olympics and Cascades. **Associates:** lily-of-the-valley, foamflower, Solomon's seal.

Another three-leaf plant carpets lowland forests of the Olympics and south Cascades. Oregon wood sorrel, or Oregon shamrock (*Oxalis oregana*) has a white-to-pinkish flower and a large cloverlike leaf trio that neatly folds up on chill nights.

LARGE SOLOMON'S SEAL

Giant false Solomon's seal, solomon-plume, false spikenard, western smilacina

Smilacina racemosa

Lily Family

The long stalks with double rows of heavily-veined leaves are showy, and so are the bursts of creamy flowers, and so are the large berries that follow, first spotted, then orange or red. As for the perfume, woods full of the bloom are as over-powering as a sultan's seraglio.

Flowering season: May to July. **Features:** leaves up to 8 inches long on a stalk up to 40 inches long, usually somewhat drooping, the flower plume at the end. **Habitat:** moist woods from lowlands to middle elevations. **Distribution:** Olympics and Cascades. **Associates:** vanilla leaf, queen's cup, foamflower, Canadian dog-wood, creeping raspberry.

The small, or star-flowered Solomon's seal (*S. stellata*) has fewer and smaller leaves; most distinctively, the flower cluster consists of only a few, perhaps just several, loose blossoms.

Similar in leaf and structure are the twisted stalks (page 34) and fairy bells, but their flowers are semiconcealed under the stem rather than boldly in the open at the tip of the stalk.

DEVIL'S CLUB
Oplopanax horridum
Ginseng Family

With such plants as this the jungles of Hell torture lost souls — and many an off-trail hiker in jungles of the Northwest has felt abundantly punished for sins of bad routefinding. No botanist's "horridum" is needed to warn against the wicked clubs armed with spines that cause cries of anguish as they stab the skin and then, being barbed, stay on to fester. However, the walker whose path lies not through but beside the treelike tangles can pause on his virtuous way to admire in peace.

Flowering season: May to July. **Features:** small greenish-white flowers, scarcely noticeable, produce large pyramids of berries, first yellow, then scarlet; large maplelike leaves with spines on underside veins; up to 10 feet tall. **Habitat:** moist forests and swamps from lowlands to middle elevations. **Distribution:** Olympics and Cascades. **Associates:** western redcedar, thimbleberry, yellow violet.

QUEEN'S CUP

Clintonia, beadlily, bluebead, alpine beauty, bride's bonnet

Clintonia uniflora

Lily Family

Just as another lily, the trillium, announces springtime in the lowlands, at higher elevations the queen's cup whitens the forest floor still wet from the recent melting of the snow.

Flowering season: May to July. **Features:** often in large groups; up to 6 inches tall; the flower matures to a striking beadlike berry, a dark, metallic blue. **Habitat:** rich soil in moist forests from middle to subalpine elevations. **Distribution:** Olympics and Cascades. **Associates:** lily-of-the-valley, Solomon's seal, Canadian dogwood.

Lily-of-the-valley, or beadruby (*Maianthemum dilatatum*) grows in large patches of conspicuous leaves, shiny and pointed. The spikes of tiny white flowers develop to pretty ruby beads.

Starflower (genus *Trientalis,* in the Primrose Family), prominent in lowland forests, has a crowd of leaves near the tip of the stem and a white, star-shaped flower.

Flowers of the Forest, White

CANADIAN DOGWOOD

Bunchberry, ground dogwood, dwarf cornel, crackerberry, pigeonberry, puddingberry

Cornus canadensis

Dogwood Family

The actual flowers, in the strict technical sense, are teeny green clusters that nobody ever notices. What everybody always notices are the large, creamy-white "bracts" (modified leaves) that turn brownish with age — and the brilliant clumps of orange-red berries that color the forest floor in late summer.

Flowering season: June to August. **Features:** forms extensive mats, usually only several inches tall; as with the tree, produces a second bloom, sparse but startling, as winter nears. **Habitat:** not-too-dry, not-too-wet forests, lowland to subalpine, most abundant at middle elevations. **Distribution:** Olympics and Cascades. **Associates:** queen's cup, twinflower, creeping raspberry.

The family resemblance of leaves and flowers is unmistakable in the Pacific dogwood (*C. nuttalli*), the most striking flowering tree of lowland forests, and the red osier dogwood (*C. stolonifera*), a shrubby tree of lowland riverbanks.

TWINFLOWER
Linnaea, longtube or western twinflower
Linnaea borealis
Honeysuckle Family

Karl von Linne (or, in the Latin favored by scientists of the 18th century, Carolus Linnaeus), the Swede who devised the binomial (two-word) classification system that is the foundation of modern botany, never was portrayed except holding a sprig of linnaea, his favorite flower, which pleases the eye and the nose as well — the subtle perfume is among the most enchanting of forest aromas.

Flowering season: June to August. **Features:** creeping vines, often in large carpets; small, shiny, evergreen leaves, in pairs; white-to-pink flowers, always in twins unless some accident has removed one. **Habitat:** somewhat dry or sunny sites in forests, lowland to subalpine. **Distribution:** Olympics and Cascades. **Associates:** salal, Oregon grape, Canadian dogwood, pipsissewa, creeping raspberry.

Twinning is characteristic of the Honeysuckle Family. Twinberry (*Lonicera involucrata*), a shrub of brushy mountain valleys, has pairs of yellow flowers that mature to black berries. Familiar in lowland woods is the orange honeysuckle (*L. ciliosa*).

Flowers of the Forest, White

WOODNYMPH

Single-flowered wintergreen, single delight, waxflower, Olaf's candlestick

Pyrola uniflora

Heath Family

Many a lovely flower the hiker becomes so accustomed to he can pass by without breaking stride. The woodnymph, though, is a boot-stopper, not really rare but not too frequently seen, either. No one, walking deep shadows of big-tree forests and spotting the solitary blossom, small and often partly hidden, can continue without a pause to examine the exquisite detail, the Grecian perfection. The various names given over the years suggest how people feel about it.

Flowering season: June to July. **Features:** oval, toothed leaves at the base of a stem up to 6 inches tall, carrying a single blossom. **Habitat:** sites where little else grows but moss, typically on rotten logs vanished into the soil, in old-growth forests from lowlands to middle elevations. **Distribution:** Olympics and Cascades. **Associates:** Indian pipe.

For other members of the *Pyrola* genus that are much more common see page 30.

CREEPING RASPBERRY
Three-leaved bramble, dwarf bramble, trailing rubus
Rubus lasiococcus
Rose Family

Creeping along the ground like a strawberry, and with a similar flower, when the fruit develops it's a raspberry all right — or rather part of one, a cluster of merely several little red globes. But good.

Flowering season: June to August. **Features:** thornless stems creeping 6 feet or more, periodically rerooting; three-lobed leaves. **Habitat:** open, dry sites in forests or brushy areas from middle to subalpine elevations. **Distribution:** Olympics and Cascades. **Associates:** Canadian dogwood, queen's cup, foamflower, vanilla leaf.

The equally common five-leaved bramble, or trailing rubus, or strawberry bramble, or trailing raspberry (*R. pedatus*) has five leaflets and the fruit consists of separate, hard-looking, red globes. For a shrubby raspberry see page 25.

In logged and burned sites at low elevations grows the familiar trailing or wild blackberry, or dewberry (*R. ursinus*); on dry slopes at middle elevations is the subalpine blackberry, or snow dewberry (*R. nivalis*), virtually identical to the eye except the vine is only several feet long. Several species of strawberry (genus *Fragaria*) grow from beaches to sunblasted ridges.

SALAL
Oregon wintergreen
Gaultheria shallon
Heath Family

English gardeners greeted salal with rapture when David Douglas introduced it there early in the 19th century. In the Northwest, though, it is robbed of due respect by the very fact of being the most common forest shrub. Even the purple-blue berries, cherished by Indians, have few fans nowadays but birds — and jelly-makers.

Flowering season: May to July. **Features:** shiny, leathery leaves and wiry stems forming knee-high or over-the-head thickets. **Habitat:** moist slopes in forests from lowlands to middle elevations. **Distribution:** Olympics and Cascades. **Associates:** Oregon grape, twinflower, pipsissewa.

Usually growing nearby is one of the three local species of mahonia, or Oregon grape (Barberry Family, *Berberis* genus), a shrub with shiny, barbed leaves and yellow flowers that produce dusty-blue "grapes" that make great jelly.

Two dwarf salals often are mistaken for puny specimens of *shallon*. Slender wintergreen, or western teaberry (*G. ovatifolia*) grows on dry sites from lowlands to subalpine elevations. Alpine wintergreen, or mountain teaberry (*G. humifusa*) has even smaller leaves and grows near timberline. Both have bright red berries.

WHITE RHODODENDRON
Mountain misery, Cascades azalea, skunkbush
Rhododendron albiflorum
Heath Family

A passing hiker enjoys the masses of creamy flowers, and even the faint whiff of skunk. But pleasure turns to misery if he must battle through the thickets, as is often necessary for off-trail explorers emerging from subalpine forest into meadow.

Flowering season: July to August. **Features:** clusters of 1-12 large flowers along the branch but not on the end; shiny leaves become yellow-speckled in fall, then perhaps orange or crimson; up to 6 feet tall. **Habitat:** moist sites, such as creeks, in forests of middle to subalpine elevations; often at meadow's edge. **Distribution:** Olympics and Cascades. **Associates:** black huckleberry, oval-leaf huckleberry, fool's huckleberry, mountain ash.

On east slopes of the Cascades, often in association with the rhododendron, grows mountain Labrador tea, or trapper's tea (*Ledum glandulosum*), a similar shrub distinguished by an aromatic eucalyptuslike odor and showy white flowers at the ends of branches. In bogs of lowlands and middle elevations grows bog Labrador tea (*L. groenlandicum*).

SINGLE-FLOWERED INDIAN PIPE

Ghost plant, ghost pipe, ghost flower, corpse plant

Monotropa uniflora

Heath Family

Is this a flower? It looks more like the corpse of a flower, the ghost of a flower. Where is the green of chlorophyll, of life? There is none, because this denizen of dark woods never sees enough sunlight to subsist on it, never converts solar energy to food in the chlorophyll process. Instead, aided by a fungus associated with its roots, it gains nourishment "saprophytically" — feeding on decayed remains of dead plants in the soil.

Flowering season: July to August. **Features:** a single flower atop the erect stem, up to 12 inches tall; flower and stem age from waxy white to black; often grows in clumps of several dozen. **Habitat:** old-growth forests, lowlands to middle elevations. **Distribution:** Olympics and Cascades.

For a more common Indian pipe and other saprophytes of the Heath Family, see page 22. For members of the Orchid Family with the same lifestyle, see coralroot, page 32. Another way to dispense partly or wholly with chlorophyll is to be a parasite, tapping the juices of living neighbors; for examples see the wintergreens, page 30, and louseworts, page 77.

MANY-FLOWERED INDIAN PIPE
Pinesap
Hypopitys monotropa
Heath Family

The most common saprophyte of Washington mountains, the many-flowered Indian pipe doesn't look all that much healthier than the single-flowered (page 21) — except when the sun strikes it, and that hardly ever happens.

Flowering season: May to July. **Features:** a tight, often curled, cluster of flowers atop the stem, up to 10 inches tall; flower and stem white to pinkish to yellow, aging to black. **Habitat:** rich soils in forests where the sun rarely shines, lowland to subalpine. **Distribution:** Olympics and Cascades. **Associates:** wintergreens, coral-root, pipsissewa.

The tallest local saprophyte, up to 3 feet, is pine-drops (*Pterospora andromeda*), with a single reddish stem, many nodding flowers, white to red. Shorter, with a red-and-white-striped stalk, is candystick (*Allotropa virgata*). Fringed pinesap (*Pleuricospora fimbriolata*) has a dense spike of yellowish flowers barely pushing above the ground. All are in the Heath Family.

Flowers of the Forest, Yellow

SKUNK CABBAGE
Yellow arum
Lysichitum americanum
Arum Family

For the first announcement of spring — and not in a whisper but a shout — look to the swamps and marshes and bogs. There, while the plant world all around is only just budding, great yellow flames of skunk cabbage leap up like galaxies of exploding suns. Actually the gaudiness is less from the flowers, little green-yellow blossoms tightly massed on the big stalk ("spadix"), than the massive yellow hood ("spathe"). Later the leaves grow — and grow — green monsters up to 4 feet long, largest of any native plant. The aroma, strong though not truly skunky, doesn't gain full power until leaves are crushed or the plant dies. However, it's always potent enough to attract insects, seen in crowds creeping around spadix and spathe.

Flowering season: February to June. **Features:** one of the unmistakable flowers that everybody knows. **Habitat:** boggy ground from lowlands to middle elevations. **Distribution:** Olympics and Cascades. **Associates:** western redcedar, red alder, trillium, yellow violet.

TALL YELLOW VIOLET
Smooth woodland violet, pioneer violet, Johnny jump up
Viola glabella
Violet Family

Violets mainly aren't — most a person sees are yellow. But that's not a complaint, they're bright little beauties. Wherever the chill of winter is yielding to warm breathings of spring the forest floor is likely to be sprinkled with their pansylike blossoms.

Flowering season: March to July. **Features:** not evergreen — leaves grow new each year; up to 12 inches tall. **Habitat:** moist woods, lowland to subalpine. **Distribution:** Olympics and Cascades. **Associates:** trillium, bleeding heart, salmonberry.

V. glabella is the commonest yellow violet (of the many native to the state) at all elevations of the Olympics and west Cascades, but two yellow-flowered evergreens (keeping leaves over the winter) also are abundant. From lowlands to middle elevations is the realm of the evergreen violet (*V. sempervirens*), with small, dull-green (but shiny when new and tiny), heart-shaped leaves. The other evergreen, the round-leaved violet (*V. orbiculata*), grows from middle to subalpine elevations. Both are low, ground-hugging plants.

For violets of the proper color see marsh violet (page 83).

SALMONBERRY

Yellowberry

Rubus spectabilis

Rose Family

In winter the briar patches of satiny tan canes dominate swampy bottoms everywhere. In spring arrive the fragile pink flowers, the bloom continuing for months, even after the fruiting starts. In early summer the eating begins of the large, raspberry-shaped fruit, yellow to orange to deep red, soft and juicy and semisweet. Some folks consider the berries bland; others, including bears, gulp them by the pound.

Flowering season: March to July. **Features:** thickets of thorny canes, often taller than an upright bear; three-part, coarsely-toothed, oval leaves. **Habitat:** moist sites from lowlands to middle elevations. **Distribution:** Olympics and Cascades (rare on east side). **Associates:** trillium, yellow violet, bleeding heart.

The thimbleberry (*R. parviflorus*) is thornless and has large white flowers, large "maple" leaves, and red berries unappealing to people but popular among birds and bears. On rockslides of the east Cascades grows the red raspberry (*R. idaeus*), so much like the cultivated shrub one may suppose it has escaped from some miner's garden. Common in lowlands is the ferociously-thorny black raspberry, or blackcap (*R. leucodermis*).

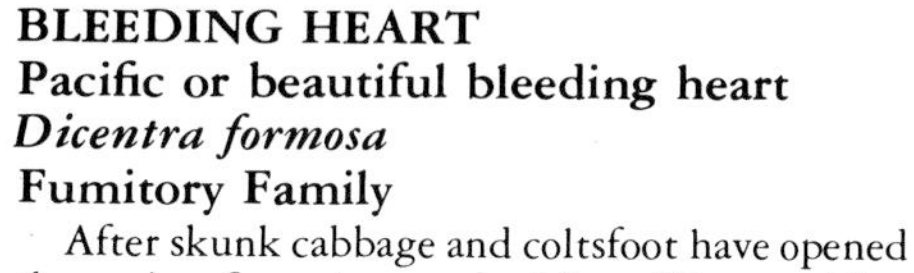

BLEEDING HEART

Pacific or beautiful bleeding heart

Dicentra formosa

Fumitory Family

After skunk cabbage and coltsfoot have opened the spring flowering, and while trillium and Indian plum and currant are carrying on the bloom, bleeding heart pushes lacy leaves through the duff, soon to bring the year's first pink tints to the forest floor.

Flowering season: March to July. **Features:** leaves finely cut, parsley-like; base of flowers heart-shaped; up to 18 inches tall. **Habitat:** shady woods, lowland to subalpine. **Distribution:** Olympics and Cascades. **Associates:** trillium, yellow violet, salmonberry.

Two other members of the Fumitory Family occur in the area. On a distant, cursory look, corydalis may be mistaken for bleeding heart. In dry, open woods up to subalpine elevations of the east Cascades grows the tiny, ground-hugging *D. uniflora,* which strikingly resembles its common name — steer's head.

WESTERN CORYDALIS
Scouler corydalis
Corydalis scouleri
Fumitory Family

Walking hour after hour through the pink of bleeding heart, becoming careless of its attraction, a person may easily pass the pink of corydalis and never note the difference, never "see" the plant at all. Actually, the color is the only real resemblance. Corydalis is far less common, but wherever there's a little there's usually a lot — whole forest slopes or entire waterfall-side cliffs solid with masses of wide-sprawling leaves from which thrust tall stalks crowded with pink flowers.

Flowering season: May to July. **Features:** always grows in dense patches; masses of flowers along the tops of stems as tall as 4 feet. **Habitat:** moist forest sites, such as river banks, mostly in lowlands. **Distribution:** Olympics and Cascades, but less and less common northward from Rainier. **Associates:** bleeding heart.

CALYPSO

Fairy slipper, Venus or lady's or angel slipper, pink slipper orchid, deer-head orchid, cytherea, hider-of-the-north

Calypso bulbosa

Orchid Family

A hiker too intent on big trees may miss this little lovely altogether, so well does it blend into the forest floor. And a hiker who delays wildland visits until high-meadow trails are snowfree also may never see it; though calypso is common in old-growth forests (but getting somewhat rare because old forests are too, and the species doesn't grow in young ones) the brief season of beauty comes and goes by early summer. For a real treat kneel and sniff. Remember what happened to Odysseus when he met Calypso.

Flowering season: May to June. **Features:** usually a single large leaf at the base of the stem, up to 5 inches tall; solitary flowers deep rose to nearly white. **Habitat:** rich soils in deep forests from lowlands to middle elevations. **Distribution:** Olympics and Cascades. **Associates:** Douglas fir, grand fir, moss.

RED HUCKLEBERRY
Vaccinium parvifolium
Heath Family

Let a big tree of the lowlands be cut or blown down and chances are in a few years the stump or snag will be topped by a cheery little garden of salal and red huckleberry — a brilliant little garden when the fruit ripens. Gourmets prefer the blue huckleberries but the reds are gratefully accepted by dry-mouthed hikers.

Flowering season: April to June. **Features:** the flowers yellow-pink cups, nodding, hiding behind leaves; up to 8 feet tall. **Habitat:** forests from lowlands to middle elevations. **Distribution:** Olympics and Cascades. **Associates:** salal.

The region's only other red huckleberry is the grouse huckleberry, or grouseberry (*V. scoparium*), a very small bush with tiny berries, common in dry upper forests and meadows of the east Cascades.

For black and blue huckleberries see page 82. All these members of the *Vaccinium* genus go under a variety of names, owing to similarities with European species. The Anglo-Saxon "wyrtil" (small "wort," or plant) evolved into "whortleberry," thence into "huckleberry." The Danish "böllebaer" (Swedish "blåbär," both meaning blue berry) evolved into "bilberry" — or in Scotland, "blaeberry."

PINK WINTERGREEN

Pink pyrola, alpine pyrola, liver-leaf or large wintergreen, shinleaf

Pyrola asarifolia

Heath Family

Trails climbing through dry midsummer forests may go for miles through wintergreens in bloom. The largest and showiest, and among the most abundant, is the pink.

Flowering season: June to September. **Features:** large, salal-like leaves at the base of the stem, up to 16 inches tall. **Habitat:** moist coniferous forests of middle elevations. **Distribution:** Olympics and Cascades. **Associates:** pipsissewa, coralroot, other wintergreens.

The green-in-winter leaves are the hallmark of the genus. The commonest is the one-sided wintergreen, or sidebells pyrola (*P. secunda*), a smaller plant than the pink, curious in that all its greenish-white flowers grow on one side of the stem. In the greenish-flowered wintergreen (*P. chlorantha*) the flowers grow all around the stem. The white-veined wintergreen (*P. picta*) has creamy flowers and distinctive leaves — dark green with prominent white veins where the plant's limited supply of chlorophyll has drained away (all pyrolas are partially parasitic).

Showing this same feature of white-striped leaves is the evergreen orchid, or rattlesnake plantain (*Goodyera oblongifolia*).

PIPSISSEWA
Western prince's pine
Chimaphila umbellata
Heath Family

The blossoms look out of place in a hot midsummer forest, appear to belong more in the showcase of the jeweler who crafted them. But it's precisely the artificial-seeming waxiness that permits them to endure heat and drought as few other flowers can. A close-to-eye inspection is required to appreciate the artful detail. Lovely — and tough.

Flowering season: July to August. **Features:** 3-8 or more nodding flowers atop a single stem up to 12 inches tall. **Habitat:** dry sites with sparse groundcover in deep forests of middle elevations. **Distribution:** Olympics and Cascades. **Associates:** wintergreens, Canadian dogwood, twinflower.

Menzie's pipsissewa, or little prince's pine (*C. menziesii*) has a shorter stem, up to 6 inches, and three or fewer flowers.

Because they are often associated and both pinkish, pipsissewa and pink wintergreen may be confused, but only on a careless glance.

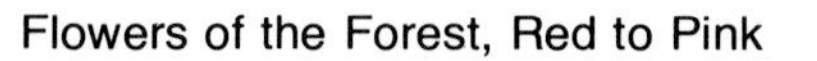

WESTERN CORALROOT
Purple coralroot, Merten's coralroot
Corallorhiza mertensiana
Orchid Family

Far back along the evolutionary track the coralroot gave up on sunlight as a direct source of nutrition, abandoned the green of chlorophyll, let its leaves shrivel to mere scales, its roots to coral-like knobs. The plant thus has an exotic, rather sinister, appearance. Yet it's not a bad sort, simply practical — in deep shadows food is more readily obtained by a saprophyte or parasite (see Indian pipe, page 21). The coralroot, a beauty of the dark, is both.

Flowering season: June to August. **Features:** entire plant, up to 20 inches tall, purplish or reddish, sometimes yellow or white. **Habitat:** brushless, deep forest from lowlands to middle elevations. **Distribution:** Olympics and Cascades. **Associates:** pipsissewa, wintergreens, Indian pipe.

Spotted coralroot (*C. maculata*) and striped (*C. striata*) are spotted or striped rather than splotched.

Flowers of the Forest, Red to Pink

COAST RHODODENDRON

Pacific or western or California rhododendron, pink or red rhododendron, California rose-bay

Rhododendron macrophyllum

Heath Family

The most famous of all Northwest outdoor shows is staged by the coast rhododendron, the region's single most gorgeous flower. Beginning in late spring along Hood Canal, the bloom climbs slopes of the Olympic Mountains, not ending there until midsummer. Even folks generally immune to wildflowers come regularly for the tour. (But not if they live in Oregon, which has more of the rhododendron — Washington's state flower — than Washington, in both the Cascades and the Coast Range. It evens up — Washington has more than Oregon of Oregon's state flower, the Oregon grape.)

Flowering season: May to July. **Features:** a nearly tree-size shrub up to 15 feet tall, sometimes forming virtual forests; large, leathery leaves. **Habitat:** open forests from lowlands to middle elevations. **Distribution:** Olympics; upper Skagit valley of North Cascades. **Associates:** evergreen huckleberry, salal, Oregon grape.

The Northwest's only other rhododendron, the white (page 20), is very different in appearance and distribution.

SMALL TWISTED STALK

Rosy twisted stalk, simple-stemmed twisted stalk

Streptopus roseus

Lily Family

A person can tramp the woods for years and never pay any attention to the twisted stalk — until, on a sidehill trail with the plant at eye level, he espies under leaves the concealed blossom bell or the translucent, bright-red berries. Or until the snows melt away in spring and he is puzzled to find the berries on the ground, still intact long after the leaves are rotted and gone.

Flowering season: June to July. **Features:** flowers greenish-yellow to purplish-rose with white tips; no twist in the "pedicel" — the linkage between blossom and stem; up to 16 inches tall. **Habitat:** damp sites in forests of middle to subalpine elevations. **Distribution:** Olympics and Cascades. **Associates:** Solomon's seal, creeping raspberry, foamflower.

The large, or white-flowered, twisted stalk (*S. amplexifolius*) is taller, grows at lower elevations, has an elongated berry, and does indeed have a sharp kink in the pedicel.

Fairy bells (genus *Disporum*) resemble the twisted stalks; the whitish bells also droop and hide but are larger and more open.

Also similar in leaf and structure, Solomon's seal (page 12) displays its flowers openly.

WILD GINGER

Asarum caudatum

Birthwort Family

Many a woods wanderer who long has admired the ginger's large, dark-green, heart-shaped leaves spreading in handsome carpets over the forest floor is surprised to learn the plant has flowers. The trick is to lift the leaves. There they are, hidden, next to the ground, odd blossoms like no other, not gaudy but fascinating in their strangeness. For more fun watch a photographer trying to prop up leaves with twigs to get a picture.

Flowering season: April to July. **Features:** large, low-growing leaf carpets, only several inches tall; a faint ginger odor, more pronounced in the underground parts. **Habitat:** rich soil in moist woods from lowlands to middle elevations. **Distribution:** Olympics and Cascades. **Associates:** vanilla leaf, foamflower, twinflower.

JACOB'S LADDER

Showy polemonium, skunkleaf polemonium, blue skunkleaf

Polemonium pulcherrimum

Phlox Family

A hiker may sit to rest in open highland woods, pleased by delicate hues of the "very handsome" (to translate *pulcherrimum*) polemonium. But (sniff, sniff) where is that indelicate odor coming from? Not a skunk but the flower, or rather the leaves, attracting flies to do the pollinating.

Flowering season: June to August. **Features:** leaves like rungs of a Jacob's ladder; up to 16 inches tall. **Habitat:** moist sites in subalpine forests and at meadow's edge. **Distribution:** Olympics and Cascades. **Associates:** lupine, penstemon, daisy, cinquefoil.

In lowlands and middle elevations occurs the giant polemonium (*P. carneum*), up to 40 inches tall. In alpine meadows is a small variety of *P. pulcherrimum,* just 2-4 inches tall. Of the same size, at the cold and rocky uppermost limit of flowering plants grows the slender or elegant polemonium, or sky pilot (*P. elegans*). From a dense tuft of leaflets rise large, gorgeous, intensely blue or purple flowers. The aroma is pungent beyond mere indelicacy and thus the other name of skunkflower. Or as some put it, polecat-emonium.

TWEEDY'S LEWISIA
Lewisia tweedyi
Purslane Family

Tweedy's is the lewisia the wildland traveler is least likely to see. Indeed, a sighting may be a never-in-a-lifetime experience unless a special trip is made, at just the right time, to one of the few sites in the very limited area where it grows. But a pilgrimage is worth the effort. Connoisseurs consider this endangered species the supreme beauty of the Northwest flower world.

Flowering season: May to July. **Features:** large flowers, up to 3½ inches wide, many on a single stem, in unique shades of pale pink, peach, or apricot; large basal cluster of thick green leaves. **Habitat:** dry, rocky, open forests, usually of Ponderosa pine, from lowlands to middle elevations. **Distribution:** east-central Cascades.

Also high on any list of great beauties is the famous bitterroot, or rock rose (*L. rediviva*), abundant and easy to see on foothills of the east Cascades. The most common lewisia is the *columbiana* (page 76).

BROAD-LEAVED MARSH MARIGOLD

White or twin-flowered marsh marigold
Caltha biflora
Buttercup Family

When subalpine meadows are squishy-soppy with meltwater from snowfields close and cold all around, there bursts forth, exuberantly ushering in springtime, the marsh marigold.

Flowering season: May to August. **Features:** long, roundish, toothed leaves at the stem base; usually two (*biflora*) flowers with 6-12 petals atop stems up to 12 inches tall. **Habitat:** beside and in snowmelt rills flowing amid or near snowbanks in subalpine forests and meadows. **Distribution:** Olympics and Cascades. **Associates:** shooting-star.

More common in some areas is the very similar heart-leaved marsh marigold (*C. leptosepala*); it has only one flower per stem and the leaves are more heart-shaped and often are rolled or folded.

In the northeast Cascades the rather rare globeflower (*Trollius laxus*) grows in the same mushy habitats and is somewhat similar, but has 5-part flowers and 5-lobed leaves that are sharp-toothed and much-dissected.

SPRING BEAUTY
Lanceleaf spring beauty, snow drops
Claytonia lanceolata
Purslane Family

When the snow melts from the meadows and spring demands to be announced, there the beauties are, often by the millions, maybe with other millions of glacier lilies.

Flowering season: April to August. **Features:** on each stem, up to 4 inches tall, a single pair of opposite "lance" leaves and several or many white-to-pinkish flowers, usually pink-veined. **Habitat:** sagebrush foothills, open forests, grassy subalpine meadows, alpine tundras. **Distribution:** Olympics and Cascades. **Associates:** glacier lily.

A pale yellow variety occurs around Glacier Peak and Mt. Baker. Close relatives in the Purslane Family, members of the *Montia* (Miners Lettuce) genus, have virtually identical flowers. Siberian miners lettuce, or western spring beauty or candyflower (*M. sibirica*), thrives from lowland forests to subalpine meadows. Miners lettuce (*M. perfoliata*) has a distinctive upper leaf, saucer-shaped with the stem poking through the middle. Streambank spring beauty (*M. parviflora*) has a creeping, vinelike stem and alternating rather than opposite stem leaves.

WESTERN ANEMONE

Mountain pasqueflower, tow-head baby, old man of the mountains, dishmop, mouse-on-a-stick

Anemone occidentalis

Buttercup Family

The flowering is brief, while snow is still melting. But all the long summer after, as other plants blossom, the tow-head babies are amid the color, waving in the breeze.

Flowering season: June to July. **Features:** flowers with 5-8 "petals" (actually, sepals); lacy, fernlike leaves; stem some 6 inches tall during bloom, continues to grow to some 24 inches for the fruiting. **Habitat:** subalpine meadows. **Distribution:** Olympics and Cascades. **Associates:** glacier lily, spring beauty, buttercup, lupine, paintbrush, valerian, cinquefoil, arnica, bistort.

On alpine ridges grow several similar but smaller anemones, most commonly the Drummond's (*A. drummondii*). In Puget Sound lowlands is the western white anemone, or windflower (*A. deltoidea*), with three oval leaves and a single large white flower.

In the Rose Family, the mountain avens (*Dryas octopetala*) has a similar flower and also a feathery-plume seedhead, but grows on rocky ridges where the thick, oval leaves form ground-hugging mats.

For other look-alikes see marsh marigold (page 38).

Flowers of the Meadow, White

Learning blossoms is one part of getting to know plants; another is learning the fruits, or seed-carriers, sometimes more spectacular. In the case of the western anemone the flower is very pretty indeed but it's the tow-head babies the hiker knows best, the tousled dishmops lasting on and on until finally beaten down by heavy wet snows of fall.

In many members of the Sunflower Family the fruit-seed demands as much attention as the flower, witness the common dandelion, whose airy globes of silky parachutes explode at a puff and fly away on the breeze. And a hiker pushing through a patch of fireweed in late summer emerges as white with down as after a night in a ripped sleeping bag.

In the Sedge Family, the tufts of cotton grass (genus *Eriophorum*) are familiar in boggy mountain meadows. Quite similar in seed, and also fond of soppy highland habitats, is western tofieldia, or lamb's lily (*Tofieldia glutinosa*), with grasslike leaves and nice little clusters of white flowers, but not truly notable until the lambs are ready for shearing.

And speaking of fruit, who can ignore the berries? For example, a meadowful of *Vaccinium deliciosum* in the September sun. . .

WHITE BOG ORCHID

Rein orchid, bog candle
Habenaria dilatata
Orchid Family

In lush crannies oozing springs, on soggy sidehills and boggy ground generally, the long stalks thrust through crowds of leaves of other plants (as in the photo here). Bend down and sniff: one expert describes the scent as a blend of cloves, vanilla, and syringa.

Flowering season: June to September. **Features:** stems 6 inches to sometimes 3 feet tall, clasped by short leaves. **Habitat:** meadow glades from middle to subalpine elevations. **Distribution:** Olympics and Cascades. **Associates:** butterwort, tofieldia, Lewis monkeyflower.

The green bog orchids (with green-to-greenish-white flowers) are much more common.

Also in the Orchid Family, hooded ladies tresses (*Spiranthes romanzoffiana*) has a resemblance on a quick look; the quite different flowers are in three spiral, neatly-braided ranks.

In the Lily Family, the mountain death camas (*Zigadenus elegans*) has short, grasslike leaves and a few green-white flowers.

Often associated with bog orchid is the fascinating butterwort (*Pinguicala vulgaris*). The blue-to-purple flower stands above flat-lying, greasy-looking, yellow-green leaves that catch and digest insects on their sticky surfaces.

OCEAN SPRAY

Arrow-wood, creambush spirea, mountain spray

Holodiscus discolor

Rose Family

Scarcely a forest opening or roadside in Western Washington lacks the broad-billowing shrub, almost as tall as a tree, that in early summer does indeed seem to have caught balls of spindrift blown from the ocean beach.

Flowering season: June to August. **Features:** undivided leaves; flower plumes fading to brown, the dried husks remaining all winter; wood so hard it was used by Indians for arrows; up to 20 feet tall. **Habitat:** dry forest openings from lowlands to middle elevations. **Distribution:** Olympics and Cascades. **Associates:** fireweed, salal, red huckleberry.

Often confused with ocean spray is another member of the Rose Family, goatsbeard (*Aruncus sylvester*). It is distinguished by not being a woody shrub, having leaves divided into toothed leaflets, and pencil-like strings of tiny white flowers atop stalks up to 6 feet tall.

GREEN HELLEBORE

Green veratrum, swamp hellebore, corn lily, Indian poke

Veratrum viride

Lily Family

The corn-tassel-like flower heads of the hellebore are striking, but less so than the large, prominently-veined leaves and the plant's overall giant size. Especially notable is the habit of sprawling over wet meadows to form dense jungles.

Flowering season: June to September. **Features:** up to 6 feet tall; yellow-green flowers in clusters. **Habitat:** moist-to-marshy subalpine meadows. **Distribution:** Olympics and Cascades. **Associates:** cow parsnip, valerian, lupine, paintbrush.

The white hellebore (*V. californicum*), similar except for whitish flowers, generally grows at lower elevations. The green is more common in the west Cascades, the white in the east.

Careless hikers thinking to enjoy a nice pot of boiled skunk cabbage (page 23) have mistakenly picked hellebore leaves instead and poisoned themselves; among the more colorful symptoms, the victim's skin and eyeballs turn yellow and through his eyes the whole world looks yellow.

Flowers of the Meadow, White

COW PARSNIP
Cow cabbage
Heracleum lanatum
Parsley Family

A veritable Hercules — or cow — of a plant, the cow parsnip towers above companions of the highlands. And of the ocean beaches as well.

Flowering season: May to August. **Features:** a single hollow stem as thick as a sausage, up to 10 feet tall; immense, three-part leaves; large, flat-topped flower heads. **Habitat:** boggy ground and streambanks from lowlands to subalpine meadows. **Distribution:** Olympics and Cascades. **Associates:** hellebore, valerian, lupine, paintbrush.

Sheer brute size usually distinguishes cow parsnip from other meadow whites, but sometimes the plant is small. Attention to leaves then readily separates it from valerian and meadow parsley (pages 46-47).

Also in the Parsley Family, also white and often large, are the angelicas (genus *Angelica*), with flower heads similar to cow parsnip on a quick glance but with leaves divided into oval leaflets.

In the Buttercup Family, mostly in forests, is false bugbane (*Trautvetteria carolinensis*), with large, maplelike leaves and a fuzzy head of small tubular flowers. Another forest Buttercup, the shrubby-looking western baneberry (*Actaea rubra*), also has a fuzzy white flower head but is most noticed for its clusters of shiny red berries.

SITKA VALERIAN
Mountain heliotrope
Valeriana sitchensis
Valerian Family

There's a lot of white in a mountain meadow, plenty of confusion for the eye of the beginning student. However, the nose quickly and surely singles out the Sitka valerian, among the chief perfumers of the highlands, so sweetly powerful that on still and sultry summer days it may very well give a hiker an attack of giddy ecstasy.

Flowering season: July to August. **Features:** flowers white to pinkish; leaves, mostly on stem, composed of usually 3 but up to 7 separate leaflets, toothed; up to 3 feet tall. **Habitat:** lush subalpine meadows. **Distribution:** Olympics and Cascades. **Associates:** meadow parsley, lupine, paintbrush, bistort, cow parsnip.

The only other valerian common in Western Washington is Scouler's (*V. scouleri*), flowering in spring in open spots from saltwater beaches to middle elevations. Though the plant is small the flowers are very pretty; however, the appeal is to the eye, with not much to excite the nose.

To distinguish valerian from other meadow whites, see meadow parsley.

MEADOW PARSLEY
Gray's lovage, licorice-root
Ligusticum grayi
Parsley Family

In lush subalpine meadows of late summer and fall, meadow parsley is often among the most prominent flowers. Where there is any of it there usually are fields of it, blended with other whites — and blues and yellows and reds.

Flowering season: July to September. **Features:** lacy "parsley" leaves, large at the stem base, sometimes a few tiny ones higher; white-to-pinkish flowers in round-topped clusters of small balls; up to 24 inches tall. **Habitat:** open subalpine woods and meadows. **Distribution:** Cascades. **Associates:** valerian, bistort, lupine, cinquefoil, paintbrush.

Canby's lovage (*L. canbyi*), up to 4 feet tall, grows in the east Cascades. At lower elevations than the meadow parsley, down to sealevel, is its look-alike relative, the water parsley (*Oenanthe sarmentosa*).

To distinguish other meadow whites: valerian has a very different leaf and a sweet perfume; cow parsnip (page 45) is grossly larger; bistort (page 48) and pearly everlasting and yarrow (page 49) are alike only in being white.

MOUNTAIN BISTORT

Mountain dock, common or American bistort, mountain meadow knotweed, snakeweed

Polygonum bistortoides

Polygonum (Buckwheat) Family

Waving in summer winds atop long stems, or sagging as bumblebees sit, the distinctive flower heads rise above surrounding reds and yellows and blues, dotting the meadows with balls of white.

Flowering season: June to September. **Features:** a dense spike of tiny flowers that last far into fall, atop a stem up to 2 feet tall. **Habitat:** the highest subalpine meadows. **Distribution:** Olympics and Cascades. **Associates:** valerian, meadow parsley, paintbrush, lupine, groundsel.

Alpine bistort (*P. viviparum*) is shorter.

Buckwheats (genus *Eriogonum*) often abound in dryish meadows, especially of the east Cascades. Typically they have dense balls of small flowers, reddish-white to brilliant yellow.

Mountain sorrel (*Oxyria dignya*), growing in rocky alpine sites, has a spike of small, reddish-tinged green flowers and fleshy leaves with a rhubarb (oxalic acid) tang.

Other common ball-heads are pussytoes (Sunflower Family, genus *Antennaria*) and pussypaws (Purslane Family, *Spraguea umbellata*), a high alpine plant with prostrate leaves upon which lie the fuzzy purple-to-white flower balls.

Flowers of the Meadow, White

PEARLY EVERLASTING
Indian tobacco
Anaphalis margaritacea
Sunflower Family

The plant is so common that in summer the casual eye passes blindly by, dazzled by other blossoms in full riot. But when surrounding colors fade the eye begins to linger and finds a genuine pearl, closely resembling its famous relative, symbol of the Alps, the edelweiss. Frontier homesteaders brightened cabin walls with it in winter; if picked in bloom the pearly indeed seems everlasting.

Flowering season: July to October. **Features:** papery white "bracts" that seem part of the flower surround the true flowers, which are yellow; narrow, long leaves; up to 3 feet tall. **Habitat:** dry, open sites from lowlands to subalpine meadows. **Distribution:** Olympics and Cascades. **Associates:** yarrow, goldenrod.

Another common "weed" in the same family, yarrow (*Achillea millefolium*), pioneers dry, poor soils and wastelands from sand dunes by ocean surf to rock barrens at the uppermost limit of flowering plants. The leaves are long ladders with little fuzzy rungs; the heads are wide, flat masses of small white flowers with yellow centers.

Another late-bloomer of dry sites at all elevations is the familiar goldenrod (genus *Solidago*).

CASCADE MOUNTAIN ASH
Sorbus scopulina
Rose Family

The transition from a clump of subalpine trees to an open meadow frequently is a copse of mountain ash, a shrub of nearly tree size, showy in early summer with creamy flower balls, in fall with masses of bright red or orange berries that migrating birds eat by the ton. So do bears; their very rapid and highly incomplete digestion of the fruit distinctively decorates autumn trails.

Flowering season: May to July. **Features:** leaves composed of ladders of glossy-green, pointed leaflets; up to 10 feet tall. **Habitat:** subalpine meadow edges. **Distribution:** Olympics and Cascades. **Associates:** white rhododendron, black huckleberry, fool's huckleberry.

More common in the Cascades is Sitka mountain ash (*S. sitchensis*), distinguished by leaflets that are a dull grayish-green and rounded at the outer end. Familiar in and near gardens is the European mountain ash (*S. avcuparia*), a tree.

The Rose Family has many other flowering shrub-trees common from lowlands to mountain valleys, including Indian plum, bitter cherry, ninebark, and serviceberry, all with white blossoms.

Flowers of the Meadow, White

BEARGRASS

Squaw grass, elk grass, Indian basket grass, bear lily, wild turkey beard

Xerophyllum tenax

Lily Family

Among the most exhilarating of flower experiences is a walk through a field of beargrass in bloom. The hugeness of the creamy-white masses makes the first impression, but there is quite another, equally strong, when a flower head is held close to the eye and seen to be composed of myriad tiny blossoms. The flowering begins as a small, dense clump that eventually expands along the stem a foot or more, the stem meanwhile continuing to grow until it reaches a height of as much as 6 feet — unless a deer gets there first and neatly clips off the stalk and eats virtually the whole plant. Beargrass usually grows where a forest has been destroyed by fire, wind, insects, or logging. The plant is thought to die after flowering — certainly, a field covered with blossoms one year may have none the next.

Flowering season: May to August. **Features:** very long, grasslike leaves from the base, a few small ones on the stem. **Habitat:** open sites mostly, but sometimes in forests, at sealevel along the ocean but mainly in subalpine woods and meadows. **Distribution:** Olympics and Cascades south of Glacier Peak. **Associates:** black huckleberry, Canadian dogwood.

WHITE PAINTBRUSH
Small-flowered Indian paintbrush
Castilleja parviflora var. *albida*
Figwort Family

The botanical classification system nearly cracks wide open when it collides with the *Castilleja* genus. Perhaps due to hybridization, and/or narrow adaptation to the chemistry of specific soils, there seems an infinite number of distinct paintbrushes. How does one organize them into species? Very carefully. By use of many subspecies or varieties, some looking quite different. Here, for instance, is a white that belongs to the same species as the magenta on page 75. Interesting.

Flowering season: July to August. **Features:** white or yellowish to pinkish flowers; leaves divided into 3-5 lobes; woody-base stem up to 12 inches tall. **Habitat:** moist subalpine meadows. **Distribution:** North Cascades (mostly northeast), mainly from Glacier Peak north. **Associates:** lupine, groundsel, valerian, meadow parsley, bistort, arnica.

The only other white paintbrush in Western Washington is the Mount Rainier paintbrush (*C. cryptantha*), limited to that vicinity, where — happily for students — the *S. parviflora* var. *albida* doesn't occur.

Flowers of the Meadow, White

WHITE HEATHER

White moss heath or heather, mountain heather

Cassiope mertensiana

Heath Family

Except in gardens the Northwest lacks the true heathers of Europe, plants in the Erica Family. Very well, so Washington heathers are really heaths. They nevertheless stir memories of Scottish highlands; perhaps it was the family connection that made this John Muir's favorite flower. For others the fondness comes simply from the masses of white bells gladdening the eye as breezes from the snow do the cheek.

Flowering season: July to September. **Features:** moss-resembling leaves composed of overlapping small scales; fleshy, berrylike fruits in late summer; up to 15 inches tall. **Habitat:** subalpine meadows. **Distribution:** Olympics and Cascades. **Associates:** red heather, yellow heather, Cascades huckleberry, lupine.

Uncommon except at Mount Rainier is the moss heather, or Alaska heath (*C. stellariana*). The white flowers are single on the ends of stems rather than in bunches near the ends as with *C. mertensiana*.

Also in this photo is red heather; for discussion of it and the yellow see page 80.

PARTRIDGEFOOT

Luetkea, Alaska or alpine or meadow spirea

Luetkea pectinata

Rose Family

The fringed, lacy leaves have the look to some eyes of the feathered feet of a grouse, called by some a partridge. The spreading carpets, the short stems with showy heads of superb little flowers, have the look to all eyes of snow country. Indeed, in cold corners where the snow stays late it's usually partridgefoot, if anything, that's there after the melt, pioneering the bare dirt.

Flowering season: June to August. **Features:** shrubby plants mostly under 4 inches tall, forming mats of mosslike foliage. **Habitat:** sandy-gravelly but moist sites from subalpine meadows to the upper limits of flowering plants. **Distribution:** Olympics and Cascades. **Associates:** valerian, paintbrush, Cascade huckleberry.

Flowers of the Meadow, White

TOLMIE SAXIFRAGE
Alpine saxifrage
Saxifraga tolmiei
Saxifrage Family

Flowers of the Saxifrage Family typically are exquisite miniatures, little pictures demanding to be examined in fine detail. Those of the Tolmie saxifrage are the more delightful for the bleakness of the high, rocky barrens the plant loves to pioneer.

Flowering season: July to September. **Features:** stubby, fleshy, stonecroplike leaves; usually less than 4 inches tall. **Habitat:** bare gravel slopes and rock crannies, mainly at the highest elevations of flowering plants. **Distribution:** Olympics and Cascades. **Associates:** as a pioneer, mostly grows alone.

Other jewels of the high alpine ridges are matted, or spotted saxifrage (*S. bronchialis*), with flowers growing from dense mats of tiny spike-tip leaves, and tufted saxifrage (*S. caespitosa*), from mats or tufts of finely-divided leaves.

Several other saxifrages are common from meadows on down, usually in wet places, notably the cool spray of waterfalls. On sunny cliffs are the airy clusters of alumroot. Much of the forest bloom is from foamflower, youth-on-age, fringecup, and mitrewort. And east-Cascade foothills are brightened in spring by the prairie star. It's a big and wonderful family.

AVALANCHE LILY

Alpine avalanche lily, alpine fawn-lily, trout-lily, adder's tongue

Erythronium montanum

Lily Family

GLACIER LILY

Yellow avalanche lily, snowlily, lambstongue, dogtooth violet

Erythronium grandiflorum

Both avalanche lily (white) and glacier lily (yellow) grow in fields, often covering whole valley bottoms or mountainsides. The flowers begin their bloom at the very edges of snowfields; not long after the snow is gone, so are they. Glacier lily has a wide elevation range, from east-Cascade sagebrush prairies where it blossoms in March to the very high meadows. Avalanche lily favors lower meadows and in the west Cascades descends some distance into forests.

Flowering season: March to August. **Features:** long, pointed leaves turning a mottled yellow-brown with age. **Habitat:** subalpine meadows and open forests moist from snowmelt. **Distribution:** Olympics and Cascades. **Associates:** spring beauty.

In Western Washington lowlands grows the spring-blooming great avalanche lily, or Easter lily (*E. oregonum*), with distinctively brown-mottled leaves and a white flower.

Flowers of the Meadow, White and Yellow

YELLOW BELL

Yellow fritillary, mission bell, yellow snowdrop

Fritillaria pudica

Lily Family

A mass display of glacier lilies is a glory of yellow at the snowfield's edge. The yellow bell, also blooming when snow is still melting all around, doesn't dazzle, it charms. Often growing alone in the austerity of dry barren or sparse meadow, the demurely-nodding blossom is easily missed. But the eye that once has spotted one keeps a sharp lookout ever after.

Flowering season: April to June. **Features:** narrow, pointed leaves on stem up to 5 inches tall. **Habitat:** dry, grassy sites from sagebrush prairies to subalpine meadows. **Distribution:** east Cascades. **Associates:** spring beauty, shootingstar, grass widow.

The brown fritillary, or chocolate or leopard lily (*F. lanceolata*), grows from sealevel to foothills in Eastern and Western Washington woods and meadows. Its nodding bowl is purplish-brown, speckled on the inside with green-yellow.

COLUMBIA TIGER LILY
Columbia lily, Oregon lily
Lilium columbianum
Lily Family

If tigers had spots this lily might resemble them, at least in brilliance. But they don't, so why the name? Perhaps because of the tigerish way the fierce blossoms leap out at a hiker from the greenery of a summer meadow. Certainly on first startled sight a person must gasp. According to English folk belief there truly is peril in the flower — sniff it and you get freckles.

Flowering season: May to August. **Features:** lance leaves in a whorl circling the stem; up to 4 feet tall. **Habitat:** openings in forests from lowlands on up, but mainly in subalpine meadows. **Distribution:** Olympics and Cascades. **Associates:** columbine, penstemon, paintbrush, Merten's bluebells.

On dry slopes in the southern part of the east Cascades another dramatic lily leaps out at the hiker. The subalpine Mariposa lily, or sego lily or cat's ear or butterfly (*Calochortus subalpinus*), has a large, creamy flower with a narrow purple crescent on each of the three petals.

SNOW BUTTERCUP
Subalpine or mountain buttercup
Ranunculus eschscholtzii
Buttercup Family

Buttercups are everywhere from lowlands to highlands, by the dozens of species, 16 in Western Washington alone. And there also are a lot of yellow look-alikes that are other things altogether — notably the cinquefoils. The snow buttercup, however, is just about unmistakable. No elbow-jostler in the summer confusion is this but a lonesome herald of spring, starting to grow while under the snow, bursting forth in bright bloom beside — and in — meltwater rills flooding meadows otherwise still winter-brown.

Flowering season: June to July. **Features:** waxy petals; fleshy, shiny, kidney-shaped leaves; often merely 2 inches tall but up to 10 inches. **Habitat:** subalpine meadows. **Distribution:** Olympics and Cascades. **Associates:** spring beauty, western anemone.

In moist-to-boggy fields and forest openings of lowlands the western buttercup (*R. occidentalis*) is common, a rank plant up to 2 feet tall, usually sprawling in large patches. Also common in moist lowland woods is the woods buttercup (*R. uncinatus*), with small blossoms usually missing some petals; in late summer little seed-carrying ballheads hook onto pants of passing walkers and hair of passing dogs.

Flowers of the Meadow, Yellow to Orange

SHRUBBY CINQUEFOIL
Five finger
Potentilla fruticosa
Rose Family

Cinquefoils contribute a lot of yellow to meadows. The climax of the genus, though, is high on dry alpine ridges. There the shrubby cinquefoil sprawls among sterile rocks, the splashes of massed bloom truly stunning.

Flowering season: July to August. **Features:** not an herb, as are other cinquefoils, but a shrub with woody stems. **Habitat:** dry sites, mostly in high subalpine meadows. **Distribution:** Olympics and Cascades. **Associates:** Davidson penstemon, phlox, harebell, yarrow.

Of the eight members of the genus in Western Washington, most common in subalpine meadows is the fan-leaf, or Rainier cinquefoil (*P. flabellifolia*), with 3 leaflets spread in a fan, and the mountain meadow cinquefoil (*P. diversifolia*), with 5-7 leaflets. Also common is the similar large avens (*Geum macrophyllum*).

Is what you are looking at a cinquefoil or a buttercup? Buttercups like cool north slopes and wet spots and have varnished-looking, unnotched petals. Cinquefoils like warm south slopes and dryish spots and have velvet-dull, notched petals. Turn the blossom over: if on the bottom there are 5 little pointy triangles, yellow, it's a buttercup; if there are 10 green ones, it's a cinquefoil.

TALL WESTERN GROUNDSEL
Senecio, butterweed, ragwort
Senecio integerrimus
Sunflower Family

A hiker ascending a trail meets groundsels in low valleys, groundsels on sky-airy ridges, the plants tall and lush in wet meadows, dwarf and pinched on barren crags. The tall western groundsel covers pretty much all elevations.

Flowering season: May to July. **Features:** oval leaves, long at the base, shorter upwards; white-to-yellow flowers in a clump of many heads atop the stem, a few inches to 2½ feet tall. **Habitat:** wet sites and dry, lowland woods to alpine ridges. **Distribution:** Cascades, mostly east side. **Associates:** balsamroot, desert parsley.

The commonest member of the genus in the west Cascades, also abundant on the east, is the spear-headed, or arrowleaf groundsel (*S. triangularis*), up to 3 feet tall, easily recognized by its large triangular (spearhead) leaf.

Of the many other members of the huge Sunflower Family that also feature "yellow daisies," particularly prominent in mountains are those in the *Arnica* genus; the most common is the subalpine arnica (*A. latifolia*).

The quick way to distinguish the two genera: groundsels have leaves alternating on opposite sides of the stem; arnicas have opposite leaves and fewer flower heads per stem, often only one.

GOLDEN DAISY
Golden fleabane, yellow aster, alpine yellow daisy
Erigeron aureus
Sunflower Family

Way high on the ridges where winds are loud and cold and flowers miniature and few, scattered through sparse tundras and over fields of stones ("fellfields") and seas of boulders ("felsenmeer"), the little golden daisy finds nooks to sustain bursts of color — usually with not many companions as brave.

Flowering season: July to August. **Features:** flower heads up to 1 inch in diameter atop stems rarely more than 6 inches tall. **Habitat:** rocky, high, and dry alpine ridges. **Distribution:** Cascades. **Associates:** golden dasiy (yellow), Lyall's lupine (blue), and alpine paintbrush (scarlet), all dwarfs and all brilliant, form a famous and spectacular trio of the high Cascades.

For other members of the *Erigeron* genus see mountain daisy (page 86).

DESERT PARSLEY

Lomatium, spring gold, hog fennel
Lomatium genus
Parsley Family

A springtime sprinkling of gold dust on rocks of a lowland desert — or of a sunny highland buttress — earn the flower one of its nicer names. But some members of the genus in some favored sites produce veritable ingots.

Flowering season: April to July. **Features:** carrotlike leaves; plant low when flowering starts, continues to grow until up to 12 inches tall. **Habitat:** dry, rocky sites from lowlands to subalpine meadows. **Distribution:** Olympics and Cascades. **Associates:** yarrow, harebell, onion, blue-eyed Mary.

The casual student doesn't distinguish the numerous desert parsleys that grow in this place and that. Yellow is the commonest color but some are white and others purple. Only two grow west of the Cascades, most abundantly *L. martindalei,* which may be either white or yellow.

SPREADING STONECROP
Cascade stonecrop, sedum
Sedum divergens
Stonecrop Family

For sure, it's a fine and lovely crop to come from a pile of stones. The bright yellow stars are the more splendid for the barrens from which they blossom, and the fleshy leaves (designed to store water for the dry season) indeed look good enough to eat. (They aren't.)

Flowering season: July to September. **Features:** bulbous, fleshy leaves that don't appear to be leaves at all — "succulent" leaves like those of the Tolmie saxifrage (page 55); seed pods that spread apart, giving the species name. **Habitat:** dry, rocky sites in subalpine meadows and to the upper limits of flowering plants on alpine ridges. **Distribution:** Olympics and Cascades. **Associates:** phlox, littleflower penstemon.

In various mountain areas, down to quite low elevations on rocky sites in forests, grow several other stonecrops, all so similar the ordinary person can't tell them apart.

TILING'S MONKEYFLOWER

Clustered monkeyflower, alpine yellow mimulus

Mimulus tilingii

Figwort Family

Some folks see a monkey, some a stage mimic, and others a dragon — snapdragon, that is. Whatever, they're happy faces, usually in mobs yellowing the gray of gravel or the green of moss, often beaded with droplets flung from a waterfall or splashing creek.

Flowering season: July to September. **Features:** typically forms a mat; from each stem, 2-4 inches tall, a single 1-inch-wide blossom that seems much too large for its little clump of tiny leaves. **Habitat:** creek gravel and beds of moss in the highest subalpine meadows. **Distribution:** Olympics and Cascades. **Associates:** Lewis monkeyflower, dwarf fireweed, alpine coltsfoot.

A half-dozen yellow monkeyflowers grow in Western Washington and adjoining mountains. Most abundant in lower subalpine meadows is the common monkeyflower (*M. guttatus*), up to 2 feet tall, with many blossoms per stem. In the high east Cascades is the very tiny primrose monkeyflower (*M. primuloides*). In lowland woods grows the also tiny chickweed monkeyflower (*M. alsinoides*). Along lowland streams, mainly in the Olympics, is the coast monkeyflower (*M. dentatus*).

LEWIS MONKEYFLOWER
Pink or red or great purple monkeyflower
Mimulus lewisii
Figwort Family

Crossing the continent with Captain Clark and company on the journey of 1804-06, Meriwether Lewis was duly impressed by the masses of rose-to-purple monkeyflowers along mountain streams. The bloom of a clump is notably long, virtually the entire summer; the first blossoms open while other buds are still developing, preparing to carry on a continuous show even as petals of predecessors fall in the water, forming rafts of color floating in pools, drifting downstream.

Flowering season: June to August. **Features:** leaves prominently veined; forms dense, sprawling clumps, often very large; usually about 1 foot tall but up to 3 feet. **Habitat:** creek banks and seepage slopes in subalpine meadows. **Distribution:** Olympics and Cascades. **Associates:** yellow monkeyflower, dwarf fireweed, alpine coltsfoot.

JEFFREY SHOOTINGSTAR

Tall mountain shootingstar, peacock, cowslip, rooster-head

Dodecatheon jeffreyi

Primrose Family

Among the first flowers to bloom in meadows still soggy with meltwater, the shootingstar continues its peacock strutting through the summer in boggy spots and beside little trickle-creeks.

Flowering season: June to August. **Features:** forms dense clumps with dozens of flowers; leaves oval to long lances; from 3 inches to 2 feet tall. **Habitat:** wet sites in subalpine meadows. **Distribution:** Olympics and Cascades. **Associates:** marsh marigold, violets.

Several other shootingstars of the same color are so similar the average person can't tell them apart by the blossom; common at lower elevations west of the Cascades is Henderson's (*D. hendersonii*), with a distinctive broad, oval leaf. In the east Cascades is the white shootingstar (*D. dentatum*).

COLUMBINE
Red columbine, Sitka columbine
Aquilegia formosa
Buttercup Family

Comely and beautiful (*formosa*) enough to compete for attention in the most flower-bright meadows, as often as not the columbine has the added color of a hummingbird stuck in its throat, sucking nectar from the long spurs of the petals.

Flowering season: June to August. **Features:** leaves split in 3 leaflets, which are split again, forming a large, deep-scalloped oval; up to 3 feet tall. **Habitat:** moist-to-dry sites in openings from lowland woods upward, but mostly in subalpine forests and meadows. **Distribution:** Olympics and Cascades. **Associates:** paintbrush, tiger lily, penstemon.

Much less common is the yellow columbine (*A. flavescens*), found mainly on the east side of the North Cascades.

Two other handsome members of the Buttercup Family occupy similar wet sites. The larkspur (genus *Delphinium*) resembles a blue columbine but has a single spur. Columbia monkshood (*Aconitum columbianum*), up to 5 feet tall, has a deep-purple-to-blue flower with the look of — yes — a monk's hood.

FIREWEED
Great willow herb, blooming Sally
Epilobium angustifolium
Evening Primrose Family

Let man or nature burn a forest anywhere in the Northwest and in following years the blackened logs and stumps will be swallowed up by fields of fireweed's tall stalks. In early summer they burst out in a vast expanse of bloom grand enough to delight any Sally. Later they loose upon the wind clouds of silky-winged seeds to be carried to other burns, or clearcuts, or avalanche swaths that need a pioneer to prepare for a new forest. To the color shows of recent clearcuts in national forests and private tree farms flock beekeepers, bringing hives to accumulate one of the most prized of honeys.

Flowering season: June to September. **Features:** lance leaves, alternate, narrow and pointed; masses of four-petal flowers along upper part of stem up to 8 feet tall; seedpods that split to release fluff balls. **Habitat:** openings in forests, lowland to subalpine. **Distribution:** Olympics and Cascades. **Associates:** thimbleberry, thistle.

Several other similar but generally smaller fireweeds, or willow herbs, grow in Washington mountains.

DWARF FIREWEED

Broad-leaved willow herb, river beauty, gravel-flat rose

Epilobium latifolium

Evening Primrose Family

A rose glow on gray gravel beside cold water rushing from the snout of a glacier, that's the river beauty. Its tall relative of lower elevations covers whole mountainsides with color, yet for splendor of individual blossoms must yield to the dwarf that lives near the ice.

Flowering season: July to August. **Features:** grows in sprawling clumps of gray-green leaves; large blossoms up to 1½ inches wide; up to 16 inches tall. **Habitat:** pioneers bare ground in subalpine meadows, river gravels at high elevations. **Distribution:** Olympics and Cascades. **Associates:** monkeyflowers, lupine, alpine coltsfoot.

Other dwarf fireweeds are similar but less common. The alpine willow herb (*E. alpinum*) is a tiny plant with a small flower.

INDIAN THISTLE
Edible thistle
Cirsium edule
Sunflower Family

The massive plant elbows neighbors aside in mountain meadows, and the prickly leaves warn hikers to give elbow room, but the flowers invite bumblebees, as shown by the photo here.

Flowering season: July to September. **Features:** flower heads are spined, woolly-white balls from which push out the pink-to-purple flower tubes; up to 7 feet tall. **Habitat:** moist meadows and open woods, lowland to subalpine. **Distribution:** Olympics and Cascades. **Associates:** penstemon, lupine, groundsel, paintbrush, valerian.

Everyone knows a thistle on sight, or touch, but knowing which thistle it is isn't so easy, partly due to general similarities of the various species and partly to their habit of hybridizing, producing baffling mongrels. The *C. edule* is the usual thistle of high Cascade meadows but *C. brevistylum,* also called Indian thistle, also grows from lowlands to the lower subalpine elevations. And in lowlands, but also in mountains traveled by horses (which eat farm hay and then deposit on trails the seed of exotic plants) are thistles from afar, including the bull or pasture thistle (*C. vulgare*) and Canada thistle (*C. canadensis*).

ROSY SPIREA

Subalpine or mountain spirea, rose-colored or pink meadowsweet

Spirea densiflora

Rose Family

An abundant and frequently broad-sprawling shrub, with large and lovely masses of flowers, the spirea nevertheless generally doesn't receive due attention. Too common, perhaps. Or outshown by more garish blooms. Yet a hiker never again will ignore the plant once he's brought his eye close to a flower head and examined its myriad individual blossoms.

Flowering season: June to August, the flower heads then lasting on as brown seed husks. **Features:** a shrub with woody stems; up to 3 feet tall. **Habitat:** wet sites along streams and dry sites on rocky hillsides in subalpine forests and meadows. **Distribution:** Olympics and Cascades. **Associates:** Davidson penstemon, shrubby cinquefoil, phlox.

S. densiflora has flat-topped, ball-like flower clusters. A generally taller shrub, mainly of lowlands, the steeplebush or hardhack (*S. douglasii*) has the same sort and color of flowers in pyramid clusters. In the east Cascades grows the white, or birch-leaved or shiny-leaf spirea (*S. betufolia*), a lower shrub with a white flower.

SCARLET GILIA
Skyrocket, foxfire
Gilia aggregata
Phlox Family

Only the most comatose traveler does not, upon first sighting this "barbaric and startling" flower, shout or howl or at least gasp "holy cow!" Walking through a field of scarlet gilia has been likened to wandering into the middle of a Fourth-of-July fireworks show. A person guilty of plucking a blossom quickly discovers another spectacular attribute — the plant takes revenge on its killer by emitting a truly revolting stench.

Flowering season: May to July. **Features:** a biennial plant, developing a rosette of basal leaves one year, a blooming stalk the second, then dying; numerous thin little leaves on a stalk about 1 foot tall, but up to 40 inches. **Habitat:** dry sites in grasslands and open woods from lowlands to subalpine meadows. **Distribution:** east Cascades. **Associates:** typically grows alone on bare ground.

MAGENTA PAINTBRUSH

Small-flowered Indian paintbrush, rosy paintbrush, painted-cup

Castilleja parviflora var. *orepola*

Figwort Family

The Indian paintbrush most seen in Washington is the *parviflora,* which has the broadest distribution, the greatest tolerance of sites, and the longest flowering season. Oddly, though, the territories of the magenta variety here and the white of the North Cascades (page 52) don't meet — the species entirely skips the Alpine Lakes.

Flowering season: June to September. **Features:** leaves cleft in 3-5 segments; up to 12 inches tall. **Habitat:** dry-to-moist subalpine meadows. **Distribution:** Cascades from Rainier south; in the Olympics the variety, also magenta, is *olympica*. **Associates:** lupine, groundsel, valerian, meadow parsley, cinquefoil.

Second most seen in Washington is the common, or scarlet Indian paintbrush (*C. miniata*), distinguished by narrow, undivided, lance leaves and flowers that range from bright crimson to soft rose but very often are orange. High on Cascades ridges is the small (4-6 inches) cliff, or alpine or flaming scarlet Indian paintbrush (*C. rupicola*). Blooming in spring on cliffs above the ocean surf is harsh Indian paintbrush (*C. hispida*), which just to confuse matters ranges upward to subalpine meadows.

COLUMBIA LEWISIA
Lewisia columbiana
Purslane Family

The breathtaking Tweedy's lewisia (page 37) is the uncontested monarch of the genus, a large flower with a small realm. And the flamboyant bitterroot rules the deserts. But the Columbia is the lewisia of mountain valleys and ridges and is no less superb than its grand relatives. A sunny rock garden beside the trail is a good place to sit for a rest and examine in fine detail the peppermint-striped blossoms.

Flowering season: May to September. **Features:** flowers white to deep pink or rose, 6-11 petals; stem about 6 inches tall from a basal tuft of narrow, fleshy leaves. **Habitat:** rock outcrops in forest openings and meadows from middle to subalpine elevations. **Distribution:** Olympics and Cascades. **Associates:** stonecrop, buckwheat.

A sharp-eyed hiker of rocky areas very high in the mountains may be lucky enough to spot the tiny dwarf lewisia (*L. pygmaea*), thought to be rare, but that may be due to a lack of sharp eyes in high places.

Flowers of the Meadow, Red to Pink

SICKLETOP LOUSEWORT

Leafy lousewort, sickletop or ramshorn pedicularis, parrotbeak

Pedicularis racemosa

Figwort Family

To be called a figwort seems shabby enough but to be additionally a lousewort surely is cruel treatment of such pretty flowers with so amusing faces. However, fair is fair, and the folk belief was that if a cow ate this "wort" (the old English word for "plant") it became lousy (the Latin for "louse" is *pediculus*), an early manifestation of something like the germ theory.

Flowering season: July to August. **Features:** flower white to pale pink, one petal curved like a sickle or parrotbeak or ramshorn; leaves toothed, long and pointed, not "ferny" as with other louseworts (pages 78-79) but more ordinary, and thus a common name of the species; leaves and stems often reddish from lack of chlorophyll — most members of this genus are partly parasitic on other plants, as are many of the Figwort Family, including the Indian paintbrushes; up to 14 inches tall. **Habitat:** open forests and meadow edges from middle to subalpine elevations. **Distribution:** Cascades. **Associates:** grouse huckleberry.

ELEPHANTHEAD

Little red elephant, pink elephants, fernleaf

Pedicularis groenlandica

Figwort Family

The genius of the louseworts is to mimic some part of a bird or animal. In this species it's the elephant and no imagination is needed to see — on each of the dozens of flowers atop each stalk — the long trunk and high brow and floppy ears.

Flowering season: July to August. **Features:** flowers in a long head atop a stem up to 2 feet tall; fernlike leaves. **Habitat:** wet subalpine meadows. **Distribution:** Olympics and Cascades. **Associates:** bog orchid, tofieldia, monkeyflower.

The six louseworts of Washington mountains are easy to tell apart. The only one without a fernlike leaf is the sickletop (page 77). The only purple-pink one besides elephanthead is the birdsbeak.

The three not pictured here, all yellowish, all grow in wet, open, subalpine sites. The bracted lousewort, or wood betony (*P. bracteosa*) is up to 4 feet tall; the greenish-yellow flower has a very short beak. The coiled (*P. contorta*) grows from 6 to 24 inches; the white-to-yellowish flower has a markedly S-coiled beak. The Rainier (*P. rainierensis*), confined to that vicinity, is no more than 12 inches tall; the yellow-to-yellowish-white flower lacks a beak.

BIRDSBEAK LOUSEWORT
Birdsbeak pedicularis, bird's bill lousewort
Pedicularis ornithorhyncha
Figwort Family

The birdsbeak is the shortest of the louseworts, with the smallest flower head, and for mimicry can't compete with pink elephants, managing only the simple beak of a nondescript bird — maybe a goose? However, in the drier, sparser meadows where neighbors also grow smaller it is less jostled than ranker cousins in boggy spots and thus gains proper attention. Many hikers consider it their favorite specifically because the plant is more demure, the blossoms not so extravagantly massed.

Flowering season: July to August. **Features:** flower beak straight; fernlike leaves often reddish from lack of chlorophyll; usually about 6 inches tall. **Habitat:** somewhat moist subalpine meadows, but drier and sparser than the lush ones favored by elephanthead. **Distribution:** Cascades from Rainier north. **Associates:** heather, lupine.

RED HEATHER
Pink mountain heather, red mountainheath
Phyllodoce empetriformis
Heath Family

From miles away a hiker in high summer may see a whole mountainside glowing pink and know the bells are blooming there by the millions, and the bees buzzing by the thousands.

Flowering season: June to August. **Features:** leaves like little needles of a fir tree. **Habitat:** subalpine forests and meadows. **Distribution:** Olympics and Cascades. **Associates:** other heathers, Cascade huckleberry, lupine.

Yellow heather (*P. glanduliflora*) has greenish-yellow, closed bells mistakable for a sickly white heather (page 53), but quickly identified by having leaves like the red. Though at certain elevations the three may mingle, generally the red dominates lower meadows, white is prominent in the middle, and yellow occupies the highest tier.

A matted shrub that resembles heather but has little black "huckleberries" is the crowberry (*Empetrum nigrum*), the fruit bitter yet edible, said to be a favorite of the Eskimos.

Flowers like those of red heather, except they are upturned saucers and grow from a tiny (inch-high) shrub in soppy meadows, characterize the alpine swamp laurel (*Kalmia microphylla*).

SPREADING PHLOX
Carpet pink, wild sweet William
Phlox diffusa
Phlox Family

A cushion of phlox in bloom, pure white or pale blue or pink or lilac, bathes the eye like a cool mountain breeze. How does it support such exuberance on dry and rocky ground of sun-baked slopes? By thrusting a tough taproot far down in sands and gravels or cracks in bedrock, halfway to the ocean, and mining water from the depths.

Flowering season: June to September. **Features:** short needle leaves on woody stems clumped in dense, low mats from which blossom a scattering of flowers — or crowds. **Distribution:** Olympics and Cascades. **Associates:** shrubby cinquefoil, pussytoes, golden daisy, harebell, Lyall lupine, littleflower penstemon.

Several other phloxes are similar. On sagebrush foothills of the east Cascades is showy phlox (*P. speciosa*), a taller and more open plant with showier flowers.

Moss campion (Pink Family, *Silene acaulis*) grows in high alpine barrens, forming dense cushions of needle leaves, the mounds topped with small pink flowers.

CASCADE HUCKLEBERRY

Blue-leaf or delicious huckleberry (blueberry, bilberry, whortleberry)

Vaccinium deliciosum

Heath Family

Leaves turned wine-red by fall frosts and backlighted by the low autumn sun give a meadow an incandescent glow, feeding the spirit of the hiker even as he is busy feeding his face, contentedly browsing the delicious blue fruit.

Flowering season: July, leading to fruit in late August. **Features:** purple berries, usually dusty-looking from a pale-blue wax coating; typically a mere several inches tall. **Habitat:** subalpine meadows. **Distribution:** Olympics and Cascades. **Associates:** heather.

In forests are species that grow as high as 6 feet: the tall blue, or oval-leaf huckleberry (*V. ovalifolium*) and Alaska (*V. alaskaense*), both with dusty-blue fruit; and the only mountain huckleberry with a dark purple fruit, most delicious of the lot, the black, or big-leaf or thin-leaf, or big or tall huckleberry (*V. membranaceum*). At low elevations, mainly on or near the Olympic Peninsula, is the evergreen huckleberry (*V. ovatum*). For red berries see page 29.

A bush similar in appearance to the tall huckleberries but with rust-colored flowers maturing to a woody, inedible fruit is fool's huckleberry, or rustyleaf (*Menziesia ferruginea*).

Flowers of the Meadow, Blue to Purple

MARSH VIOLET
Blue swamp violet
Viola palustris
Violet Family

Most violets a person meets are yellow (page 24). The marsh violet usually deserves the name but may not be seen at all, hiding as it does amid other plants or under bushes.

Flowering season: May to July. **Features:** flowers white with purple veins, or lavender, pale violet, or deep lilac; 2-10 inches tall. **Habitat:** bogs and streambanks from mountain valleys on up, but mostly wet subalpine meadows. **Distribution:** Olympics and Cascades. **Associates:** shootingstar, elephanthead, tofieldia.

The western long-spurred violet (*V. adunca*) grows from lowlands to subalpine meadows, in the latter most abundantly, and most commonly in the east Cascades. It is usually bluer, prefers drier sites, and has a prominent spur from the flower base. The small white violet (*V. macloskeyi*) hides in grass of wet ground. The rare Flett's violet (*V. flettii*) is limited to the Olympics, causing some students to wonder how a thing so lovely came into being at all with such a restricted area in which to grow.

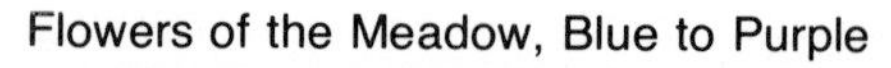

DAVIDSON'S PENSTEMON

Beardtongue

Penstemon davidsonii

Figwort Family

With gorgeous masses of outsize lavender flowers, the Davidson's penstemon could hold its own in any floral competition. The display is the more startling for coming typically from a crack in a rock wall supporting naught else but lichen, or from sere gravels of a ridge blistered by the sun.

Flowering season: June to August. **Features:** a shrubby plant with woody stems in dense, creeping mats usually lower than 14 inches; round, leathery, evergreen leaves. **Habitat:** rockslides and cliffs from middle elevations to alpine ridges. **Distribution:** Olympics and Cascades. **Associates:** shrubby cinquefoil, golden daisy, pussytoes, Lyall lupine.

Similar, at lower elevations, is the penstemon specifically called shrubby (*P. fruticosa*). Davidson's is the most common of the several evergreen shrubs; perhaps next most common is the rock, or cliff penstemon (*P. rupicola*), with thick, grayish leaves and bright rose or scarlet or lavender flowers. For nonshrubby penstemons see the littleflower.

Flowers of the Meadow, Blue to Purple

LITTLEFLOWER PENSTEMON
Small-flowered penstemon
Penstemon procerus var. *tolmiei*
Figwort Family

Blossoms of this small alpine variety of *P. procerus* are petite but by no means overwhelmed. Indeed, amid equally Lilliputian neighbors they are outstanding citizens of the sparser meadows.

Flowering season: July to August. **Features:** flowers ½ inch long or less, in one or two massed whorls atop stems up to 5 inches tall. **Habitat:** dry subalpine meadows. **Distribution:** Olympics and Cascades. **Associates:** lupine, alpine aster, phlox, yarrow, harebell.

So many are the penstemons, often so similar, the average person distinguishes only a few. Several tall ones, up to 2 feet or more, are common in forests and the lower meadows: tall penstemon (*P. procerus* var. *procerus*), identical to the littleflower but Brobdingnagian; Cascade, or spreading (*P. serrulatus*); woodland (*Nothochelone nemorosa*). For shrubby penstemons see Davidson's.

In the Mint Family are two common look-alikes, both with two-lipped flowers, the lips wider open than in a penstemon. Cooley's hedge nettle (*Stachys cooleyae*), forming "hedges" up to 5 feet tall, has red-to-purple flowers. Selfheal (*Prunella vulgaris*), 4-10 inches tall, blue to violet, is a "weed" brought into the wilds by horses.

MOUNTAIN DAISY

Common or wandering or subalpine daisy, lavender or aster or tall purple or peregrine fleabane

Erigeron peregrinus

Sunflower Family

The dandelion and the daisy — who needs an introduction to these old neighbors, familiar since childhood? Of course, most "dandelions," including all those in the mountains, are really something else. And maybe half the "daisies" are actually asters. To distinguish them, or try, see the facing page; many a person begs the question by lumping the look-alikes together as "aster-daisies."

Flowering season: July to August. **Features:** composite flower head with yellow center, white-to-lavender "petals," 30-40 in number; normally a single head per stem; up to 2 feet tall; abundant. **Habitat:** lush subalpine meadows. **Distribution:** Olympics and Cascades. **Associates:** lupine, bistort, paintbrush, valerian, cinquefoil, meadow parsley.

At high elevations three daisies are prominent. An alpine variety of *E. peregrinus* is less than 8 inches tall. The dwarf mountain daisy (*E. compositus* var. *trifidis*), 2-4 inches tall, from a clump of 3-clefted basal leaflets, is a miniature of the very high alpine barrens. Also up there is the golden daisy (page 63).

ALPINE ASTER

Dwarf purple aster, Michaelmas daisy

Aster alpigenus

Sunflower Family

The most prominent "daisy" of sparse meadows on windy ridges often is the alpine aster, seeming an old friend from city gardens — dwarfed by highland severities.

Flowering season: July to August. **Features:** most leaves from the base; only a single petite flower head per stem, with 12-20 "petals"; typically less than 6 inches tall, though sometimes up to 15 inches. **Habitat:** wet subalpine meadows to dry alpine ridges. **Distribution:** Olympics and Cascades. **Associates:** phlox, littleflower penstemon, speedwell, golden daisy, lupine.

From meadows down to lowlands are several asters much more abundant than the alpine, distinguished from it by being taller, up to 4 feet, and having leaves all the way up the stems and many flower heads per stem. Close study is required to separate these species.

As for telling daisies from asters, Mary Fries has said, "The best way is to let the experts identify them for you." Her advice is to compare photos of each, noting differences in number and shape of "petals" and attachment of head to stem, and then compare photos with the plant in question.

SUBALPINE LUPINE
Bluebonnets
Lupinus latifolius* var. *subalpinus
Pea Family

When lupine fields are blooming and summer breezes are pure perfume, a hiker may feel himself slipping away in a drugged trance, as did Ulysses and his men among the lotus eaters, bewitched in "a land that seems always afternoon."

Flowering season: July to August, then forming "pea pods" of seeds. **Features:** flowers blue to lavender, sometimes pure white; up to 2 feet tall. **Habitat:** moist sites in open forests and meadows of middle to subalpine elevations. **Distribution:** Olympics and Cascades. **Associates:** paintbrush, valerian, meadow parsley, cinquefoil, groundsel.

Broad-leaved lupine (*L. latifolius* var. *latifolius*), the same species but up to 4 feet tall, grows at lower elevations. Washington has a dozen other species, including the seashore lupine (*L. littoralis*) beside saltwater beaches. At the other altitude extreme, forming a famous trio on high Cascade ridges with golden daisy (page 63) and cliff paintbrush (page 75) is the Lyall lupine (*L. lepidus* var. *lobbi*), only several inches tall, with silver-hairy leaves.

Other genera of the Pea Family encountered in mountains are clovers, vetches, lotuses, and locoweeds.

Flowers of the Meadow, Blue to Purple

MERTEN'S BLUEBELLS

Tall mertensia, lungwort, pan bluebells

Mertensia paniculata

Borage Family

The limp, drooping flowers would be lost in the lush little jungles of stems and leaves were it not for sheer numbers, the blue bells densely crowding together. From a distance a mertensia mass is recognizable by the blue-green look of the leaves.

Flowering season: May to August. **Features:** flowers pink in bud, aging to blue; thickets up to 4 feet tall. **Habitat:** very wet sites and streambanks in upper forests and subalpine meadows. **Distribution:** Olympics and Cascades. **Associates:** monkeyflower, waterleaf.

At lower elevations are other mertensias indistinguishable to the average eye.

Also in the Borage Family are forget-me-nots, a name applied to several plants with small, pale-blue blossoms, yellow in the center. Mouse ears (*Myosotis* genus), the flower that was embroidered on garments of followers of Henry of Lancaster, is widespread. Stickseed (*Hackelia* genus) develops barbed seeds, little balls that stick to a hiker's socks and pants.

HAREBELL
Round-leaved bluebell, bellflower, lady's thimble, bluebells-of-Scotland
Campanula rotundifolia
Harebell Family

Hear the bagpipes in the blaeberry fields? This indeed is the bluebell of the Scottish highlands, equally at home in Washington, from sealevel to the sky. Peer into a bell on a sunny day to investigate the mysterious buzz and you may pull back your nose in a hurry — with a bumblebee on it.

Flowering season: May to October. **Features:** flower usually blue, sometimes white; mostly under 12 inches tall but up to 30 inches. **Habitat:** dry to moist sites in openings in lowland forests, to dry subalpine meadows, to rocky alpine ridges. **Distribution:** Olympics and Cascades. **Associates:** littleflower and Davidson's penstemon, shrubby cinquefoil, yarrow, phlox, goldenrod.

At low to middle elevations is a smaller, less showy cousin, the flower more a star than a bell, the Scouler's bluebell (*C. scouleri*). For harebells of high alpine ridges see Piper's bellflower.

PIPER'S BELLFLOWER
Piper's bluebell
Campanula piperi
Harebell Family

A person must spend a good bit of time in the alpine Olympics, and do some searching, and have some luck, to spot the rare and lovely — and due to a very restricted growing area, endangered — Piper's bellflower. In the Olympics, a mountain island cut off from other ranges, evolution has taken the opportunity to try a number of unique sideroads. The bellflower is just one species found there and nowhere else; among others are the also rare Flett's violet (page 83) and Webster's senecio (*Senecio neowebsteri*).

Flowering season: July to August. **Features:** small flower, more a star than a bell; up to 4 inches tall. **Habitat:** dry, rocky sites in subalpine meadows and on alpine ridges. **Distribution:** Olympics. **Associates:** lichens.

Far more abundant and widespread is the harebell, which has several smaller look-alikes in the Cascades: Parry's bluebell (*C. parryi*) in subalpine meadows of the central and North Cascades; rough bluebell (*C. scabrella*), a tiny, clustered plant of rocky habitats; and Alaska bluebell (*C. lasiocarpa*), 2-4 inches high, in alpine heights of the North Cascades.

ALPINE SPEEDWELL
Veronica
Veronica wormskjoldii
Figwort Family

St. Veronica wiped the drops of agony from the face of Christ on the road to Calvary, and her flowers are considered to have miraculous powers to cure scrofula. Hikers not so afflicted nevertheless can admire the sweet saintliness of visage. It's a plant that indeed seems to wish a person may "speedwell" (an old expression, the equivalent of "God be with you").

Flowering season: July to August. **Features:** oval leaves in opposite pairs; flowers deep blue to white, with four petals, one distinctly smaller, and protruding centerparts; up to 6 inches tall. **Habitat:** wet subalpine meadows. **Distribution:** Olympics and Cascades. **Associates:** lupine, littleflower penstemon, alpine aster.

Too similar to readily distinguish, but extending higher, is Cusick speedwell (*V. cusickii*). Of several species growing downwards to lowlands, most common is American speedwell, or brooklime (*V. americana*), taller and scragglier.

Another little charmer of the Figwort Family is blue-eyed Mary, or innocence or blue-lips (genus *Collinsia*), often covering bare ground beside the trail with masses of tiny two-lipped blossoms, the two petals of the lower lip usually a dark blue, the two of the upper lip lighter or even white.

MOUNTAIN BOG GENTIAN

Pleated or blue or explorer's gentian

Gentiana calycosa

Primrose Family

The last flower of summer, and of a color proper for mourning the plant world's descent into dormancy. When a hiker sees the thick buds forming, the blue petals peeping out, he knows the snows soon will blow over the highlands; it's time to eat up all the huckleberries and watch the meadows turn yellow and red and begin thinking of skis. Or beaches.

Flowering season: August to September. **Features:** flower a deep cup that closes in rainy weather; petals usually a deep blue, but sometimes entire patches are lavender or yellow. **Habitat:** moist subalpine meadows and streambanks and high alpine tundras. **Distribution:** Olympics and Cascades. **Associates:** normally grows alone, in large expanses unmixed with other flowers.

INDEX TO PICTURED FLOWERS

PHOTOGRAPHING WILDFLOWERS

Taking pictures of flowers is easy nowadays, what with modern 35mm single-lens reflex cameras, close-up lenses, and excellent color film. All a person needs is appreciation of the subject, patience to find the best angle, and a willingness to kneel to the level of the blossom. Sunshine is nice — but some of the most interesting pictures have been taken in the rain under a clear plastic umbrella.

For close-ups a camera with a macro lens is excellent but costly; inexpensive attachments are available for most cameras that work nearly as well.

At any exposure under 1/100th of a second the camera should rest on something solid. For ground-level pictures a handy rock is good, or a child's beanbag. For higher angles a sturdy tripod is best.

The major problem is wind — even a breeze the cheek can barely feel sets blossoms dancing. High-speed color film and a fast shutter speed thus are recommended. Working early in the day often helps; winds frequently are thermals caused by heat of the sun and don't start until well along in morning. Sometimes if the camera is set up and patience exerted it will be rewarded by a momentary lull — stay alert and be quick on the trigger.

Do not be guilty of trampling a hundred flowers to photograph one. Were all the thousands of photographers at Mount Rainier or Hurricane Ridge to crush even a single plant apiece, soon there would be no flower fields. So, hike far from the crowds to lonesome meadows. And then be careful where you put your feet — and your knees.

Ira Spring